Rethink

Steven Poole is the award-winning author of four critically acclaimed works of non-fiction: *Trigger Happy, Unspeak, You Aren't What You Eat,* and *Who Touched Base In My Thought Shower?* He has written widely on ideas, culture, language and society for the *Guardian*, the *New Statesman*, the *Times Literary Supplement*, the *Wall Street Journal* and many other publications.

Praise for *Rethink*

'An always entertaining and often eye-opening taxonomy of old ideas that refuse to die . . . Inspired.'

Financial Times

'Steven Poole's book is full of fascinating stories . . . It's a treasure trove for people who like to begin stories with "Did you know that . . . ?"'

The Times

'In this entertaining and important book, Poole offers a modern take on that ancient wisdom: "There is no new thing under the sun."'

Daily Telegraph

'When it comes to describing a complex idea clearly, Poole is one of the best writers around . . . This is a thoughtful and thought-provoking book.'

Sunday Times

'Engaging and enlightening . . . [A] thought-provoking book.'

Wall Street Journal

'An anecdote-rich tour through the centuries.'

New York Times

'*Rethink* refutes the "widespread assumption that it is necessary to start afresh with our uniquely modern wisdom" by illustrating the fecundity of earlier concepts and practices . . . Poole has enlightening things to say.'

Times Literary Supplement

'Shows what we can learn by considering obsolete ideas from a new perspective.'

Guardian Science Podcast

'More than a compendium of anecdotes about the forerunners of the Tesla car or the sideways history of Viagra, Poole's book is a jog on how to think . . . [An] enjoyable look at the care and feeding of creativity.'

Kirkus Reviews

'A fascinating compendium of new ideas that aren't new at all.'

New Statesman

'Poole covers a remarkable amount of ground in the history of Western thought, from ancient Greek philosophy to modern warfare . . . Such a cross-section of material guarantees there is something here for everyone.'

Publishers Weekly

RETHINK

The Surprising History of New Ideas

STEVEN POOLE

1 3 5 7 9 10 8 6 4 2

Random House Books
20 Vauxhall Bridge Road
London SW1V 2SA

Random House Books is part of the Penguin Random House group of companies
whose addresses can be found at global.penguinrandomhouse.com.

First published in paperback by Random House Books in 2017
First published by Random House Books in 2016

www.penguin.co.uk

A CIP catalogue record for this book is available from the British Library.

ISBN 9781847947581

Printed and bound by Clays Ltd, St Ives plc

Contents

1

Introduction: The Age of Rediscovery

Rethink, *v.:*
i) To think about an idea again;
ii) To change how you think about it.

'An invasion of armies can be resisted, but not an idea whose time has come.' – Victor Hugo

The electric car is the future. And it has been the future for a very long time. The first known electric car was built in 1837 by an Aberdeen chemist named Robert Davidson. It is now all but forgotten that by the end of the nineteenth century a fleet of electric taxis – known as 'hummingbirds' for their characteristic engine sound – worked the streets of London. The Commissioner of the Metropolitan Police approved of their potential to solve the city's burgeoning traffic problem, since they took up less than half the road space of horse-drawn cabs. Similar taxis also touted for trade in Paris, Berlin, and New York; by the turn of the century, more than thirty thousand electric cars were registered in the US. They were much more popular than gasoline-powered cars. They were less noisy and had no polluting exhaust. The twentieth century was obviously going to be the electric century.[1]

Yet in a little more than a decade, production of such vehicles had slowed, and then eventually stopped. The drivers of London's horse-drawn cabs had mounted a vigorous campaign pointing out breakdowns and accidents in their electrically powered rivals, and so put the London Electric Cab Company out of business.[2] (The electric taxis did suffer technical problems, but they were exaggerated by their rivals, just as the taxi drivers of modern Manhattan and London are keen to paint Uber in as bad a light as possible.) Meanwhile, discoveries of large oil reserves brought the price of petroleum tumbling, and Henry Ford began to sell gasoline-powered cars that were half the price of electric ones. The construction of better roads in America encouraged longer journeys, which electric cars could not manage on their limited battery power. So it was the internal-combustion engine, after all, that won the century.

And then, at last, came the tech entrepreneur Elon Musk, who made a fortune from co-founding PayPal and then ploughed it all into building elaborate machines in California. In 2004 he became an early funder and chairman of a Silicon Valley start-up called Tesla Motors, when almost everyone still thought that electric cars were a bad idea. 'It is frequently forgotten in hindsight that people thought this was the shittiest business opportunity on the planet,' Tesla co-founder J. B. Straubel remembers. 'The venture capitalists were all running for the hills.'[3] But Musk was able to act as his own venture capitalist. He soon became Tesla's CEO, and in 2008 Tesla launched the first highway-capable electric car, the $109,000 Roadster. It ran on lithium-ion batteries, of a similar kind to those used in laptops and phones, and it could go more than two hundred miles between charges. Most importantly, it didn't look like a clunky eco-vehicle; it looked like a flash sports car. Musk had delayed the car's development by insisting that Tesla's first model should have a carbon-fibre body and be able to accelerate from zero to sixty in less than four seconds. In doing so, he made the electric car desirable. It was a status symbol for the eco-savvy wealthy. George Clooney, Matt Damon, and Google's Larry Page and Sergey Brin bought themselves Roadsters.[4]

Tesla's next product was a slightly more sober-looking car for the mainstream market, the Model S. The S was short for sedan or saloon, but it also had a hidden historical message. Henry Ford, famously, made a Model T; well, just as S comes before T in the alphabet, so electric cars came before Ford's gasoline cars. 'And in a way we're coming full circle,' Musk told his biographer, 'and the thing that preceded the Model T is now going into production in the twenty-first century.'[5] The Model S was a hit: it was also the safest car ever tested by the American highway safety authorities.[6] By 2015 Tesla was selling fifty thousand cars per year. In the mean time established car companies such as Nissan and BMW had begun producing electric vehicles too. In 2016, Tesla announced its Model 3, with a base price of only $35,000. Within twenty-four hours the company had taken preorders worth more than seven billion dollars. 'Future of electric cars looking bright!' Musk exclaimed on Twitter. Maybe the second time around, the idea would stick.

The modern electric car is a great idea, made more viable by new technology – but it is not a new idea. And what is true in the consumer tech industry is true in science and other fields of thinking. The story of human understanding is not a gradual, stately accumulation of facts, a smooth transition from ignorance to knowledge. It's more exciting than that: a wild roller-coaster ride full of loops and switchbacks. We tend to think of the past as less intellectually evolved than the present, and in many respects it is. Yet what if the past contained not only muddle and error but also startling truths that were never appreciated at the time? Well, it turns out that it does.

This book is about ideas whose time has come. They were born hundreds or thousands of years ago. But their time is now. Many of them spent a lot of time being ridiculed or suppressed, until someone saw them in a new light. They are coming back at the cutting edge of modern technology, biology, cosmology, political thought, business theory, philosophy, and many other fields. They are being rediscovered, and upgraded. Thought of again, and thought about in new ways – rethought. Creativity is often defined as the ability to combine existing ideas from different fields. But it can also be the imaginative power of realising that a single overlooked idea has something to it after all. We are living in an age of innovation. But it is also an age of rediscovery. Because surprisingly often, it turns out, innovation depends on old ideas.

Just in time

Old is the new new. Many personal trainers have abandoned weight machines and now recommend old-school exercises such as gymnastics and kettlebells. In the food industry there is a trend for cooking elaborate feasts that the aristocracy might have enjoyed five hundred years ago. In the age of music streamed over the internet, the coolest way to release your new album is on vinyl. Bicycle-powered rickshaws ply the streets of Manhattan and London. Adults buy colouring-in books. Even airships are back. (The British-made, helium-filled, 100-metre-long Airlander 10 is designed to compete with helicopters for moving heavy cargo; others foresee a return to passenger airships in our skies, and NASA has developed concepts for airships that could resupply a space station floating in the clouds of Venus.)

And yet, alongside this multifaceted cultural turn to the past, there exists a tremendous focus on the new. Smartphones, smartwatches, fitness trackers; start-up culture and a new global skyscraper race; Uber and WhatsApp – the pace of change, we are routinely told, is greater than ever before in human history. The past is just one long roll call of mistakes. We now know better. History is old hat; the future is just around the corner. 'Conformity to old ideas is lethal,' says a former executive editor of *Time* magazine. This is 'the age of the unthinkable'.[7]

But these opinions about our culture – the retro and the futuristic, 'old is the new new' versus 'the age of innovation' – are themselves very old. And so is the tension that arises from holding them at the same time. The pace of technological change today is impressive, but hardly more so than that of the nineteenth century. The twenty-five years between 1875 and 1900 saw the invention of the refrigerator, the telephone, the electric light bulb, the automobile, and the wireless telegraph. (Not to mention the paper clip, just squeezing into the century in 1899.) Yet, in the same era, the Arts and Crafts movement was pushing a decidedly romantic return to older ideas of traditional craftsmanship and design; poets were reworking Arthurian myth; and the Renaissance was being rediscovered and hailed as the crucible of modernity. The late nineteenth century was looking both forwards and backwards, in ways that seemed unprecedented then as well.

Perhaps every age imagines itself as having a uniquely complex relationship to the past, and fails to recognise that every previous era – at least since, say, the Renaissance itself – has done so too. But what are the consequences in our day when we fail to recognise an idea as an old one? The widespread assumption that it is necessary to start afresh with our uniquely modern wisdom is encapsulated in the 'Silicon Valley ideology', which insists that venerable social institutions (such as higher education) positively need to be 'disrupted' by technology companies. The concept of innovation here is reduced to a curiously thin idea: a picture of the maverick young entrepreneur having a flash of inspiration and inventing something from nothing, to change the world. The old ways of doing things must everywhere be replaced. It is the philosophy of Snapchat, the photo and video messaging app whose messages are deleted permanently after a few seconds. What is past cannot help us; it must vanish completely.

This Valley dream of disruptive invention was beautifully satirised by a one-day event at New York University in 2014 called the 'Stupid Shit No One Needs and Terrible Ideas Hackathon', entries in which included a Google Glass app to make the user vomit on command, and a version of Tinder – for babies. If no one's had an idea before, that might be because it's an unprecedented stroke of genius – but it also might be because the idea's just not worth having. Elon Musk, for one, does not consider himself to be in the business of disruption. 'I'm often introduced on stage as someone who likes to disrupt,' he has said. 'And then the first thing I have to say is, "Wait, I don't actually like to disrupt, that sounds . . . disruptive!" I'm much more inclined to say, "How can we make things better?"'[8]

And as we'll see, innovators can often make things better by resurrecting and improving the past – as with the Tesla electric car. The philosopher of science Paul Feyerabend observed: 'No invention is ever made in isolation.'[9] Either isolated from other thinking people, or isolated in time. What other forgotten ideas might one day be rediscovered, with the help of a little rethink?

We all love a good idea, but how can we tell whether an idea is good or not? Is an idea good because it's useful, or because it will be financially profitable, or because it is morally praiseworthy, or because it inspires other thinkers? Is it an idea that helps others? Or is it an idea that merely revolutionises one's picture of the universe? Potentially any or all of the above, it seems. It is appealing to judge ideas primarily on their usefulness, but usefulness can be more narrowly or broadly conceived. Useful for what? To whom? And when? Dividing ideas into good and bad is a very blunt approach. We can do better.

For one thing, our perception of an idea will change over time. And that drives rediscovery. The electric car was a good idea given the problems of congestion and (organic) pollution caused by horse-drawn carriages. But it was arguably not such a good idea given what cheap gasoline-powered cars could do. The first electric cars could run only about thirty miles on a single charge, and in the early twentieth century there was no pressing reason for society to wean itself off fossil fuels. Advances in battery technology, together with modern climate science, have turned a problematic idea into a very good one.

So what is an idea, exactly? Is it the same as a thought, or a proposition? Is it an initial inspiration or a final conclusion? Is it a spark of genius or the outcome of long, pedantic toil? We might not be able to pin down a precise definition of 'idea', but, as the judge said about pornography, we know it when we see it. What counts as an idea is itself a subject for rethinking. And if we don't rethink the way we think about ideas, we might miss out on extraordinary possibilities.

Everybody knows, though, that some past ideas were obviously *bad*. They were just plain wrong; they were permanently superseded by new discoveries. Now we look on them fondly as just funny mistakes. Of all the discarded ideas in history, perhaps one of the least reputable is alchemy. The notion that you could turn base metals into gold? Ridiculous, of course. We regret that even Isaac Newton, for all his brilliance, was seduced by alchemical researches. To modern eyes, alchemy looks like wishful thinking at best, and esoteric fraud at worst. Alchemy is what people did before they discovered science.

Take the idea of the 'Philosopher's Tree'. According to obscure hints and scraps from laboratory notebooks of the seventeenth century, this was a precursor to the more famous Philosopher's Stone, which would turn lead into gold. By planting a specially prepared seed of gold, so it was written, you could grow a whole tree of gold. That was the Philosopher's Tree. A pretty fiction.

Except that, a decade ago, Lawrence Principe, an American chemist and historian of science, decided to see if it actually worked. He cooked up some Philosophical Mercury, a special form of mercury for which he found instructions in the secret alchemical writings of Robert Boyle, one of the founders of modern chemistry. Following a recipe that he had reconstructed from seventeenth-century alchemical treatises and experimental notebooks, Principe mixed the Philosophical Mercury with gold and sealed the mixture in a glass egg. As he watched, it began to bubble, and then rise like a baking loaf. Then it liquidised. After several days' more heating, it had turned into what Principe called a 'dendritic fractal', a structure with ramifying branches. There, before his eyes, was a golden tree.[10]

Alchemy wasn't so ridiculous after all. Historians now argue, for example, that Robert Boyle essentially 'pillaged' the work of alchemists for valuable insights while denouncing the practice as nonsense in public. In other words, he was performing a kind of dishonest rethink, taking ideas from the past and presenting them as new, while ridiculing the earlier thinkers who had them in the first place. Present-day studies have also confirmed the usefulness of ancient alchemical recipes for pigments and oils. Much of the occult weirdness, it turns out, fades away when investigators manage to translate coded terms in the old texts: according to a 2015 study in *Chemical & Engineering News*, for example, 'Historians have now figured out that dragon's blood refers to mercury sulfide, and "igniting the black dragon" likely means igniting finely powdered lead.'[11] Alchemy was not anti-science superstition; it was the best science anyone could do at the time.

When an old idea that was so obviously wrong turns out to have been right, we may be forced to rethink our ideas about the past – and our ideas about ideas themselves.

20/20 hindsight

Elon Musk's electric-car company is named after Nikola Tesla, the wizardly Serbian-American engineer and inventor active in the late nineteenth and early twentieth centuries, who pioneered the modern alternating-current (AC) electricity supply against Thomas Edison's direct current (DC). (Tesla was hauntingly impersonated by none other than David Bowie in Christopher Nolan's 2006 film *The Prestige*.) In 1888, Tesla patented a design for the first AC induction motor – the kind that Tesla Motors engineers designed their first car around more than a century later.

In 1926, Nikola Tesla was asked to imagine how the world might look fifty years in the future. 'When wireless is perfectly applied the whole Earth will be converted into a huge brain,' he declared. 'Through television and telephony we shall see and hear one another as perfectly as though we were face to face, despite intervening distances of thousands of miles; and the instruments through which we shall be able to do this will be amazingly simple compared with our present telephone. A man will be able to carry one in his vest pocket.'[12]

Wow. Smartphones.

Tesla also predicted that 'Aircraft will travel the skies, unmanned, driven and guided by radio'.

Yep: drones.

And, Tesla foresaw, 'International boundaries will be largely obliterated and a great step will be made toward the unification and harmonious existence of the various races inhabiting the globe.'

Oh well. Two out of three ain't bad.

Tesla's vision of the technology of the future was impressively accurate. But this book isn't about amazing predictions. Once you start looking, it can be tempting to see anticipations of exciting modern ideas everywhere in the past, but we must be careful in our identifications. To conclude complacently that there is nothing new under the sun is to insult those thinkers who really did discover or dream up something unprecedented. And on the other hand, to celebrate every possible foreshadowing of a good idea as brilliance would be to risk seeing genius where often there is mere chance at work. This point was well expressed by the nineteenth-century German scientist Hermann von Helmholtz. He complained about the number of untested scientific hypotheses in the journals of his time: 'Among the great number of such ideas,' he pointed out, 'there must be some which are ultimately found to be partially or wholly correct; it would be a stroke of skill *always* to guess falsely. In such a happy chance a man can loudly claim his priority for the discovery; if otherwise, a lucky oblivion conceals the false conclusions.'[13] Very true. So if someone anticipates a later idea just by guessing randomly, that's nice. But even roulette players betting on red or black are correct about half the time. So mere predictions won't constitute the kind of heroic foresight we will celebrate in those thinkers who really got something right.

Science, as we'll see, is both a field in which a lot of rethinking is currently taking place, and also itself a tool – maybe the best one we have – for rethinking. But it's a tool whose limitations are too often denied. With the best of intentions, the defenders of science in our time paint a misleading picture of it. This is perhaps understandable, since more than 40 per cent of Americans believe that God created humans a mere ten thousand years ago, and the robust findings of climate science are routinely rejected by cynical political operators. As a result, science tends to be portrayed by its advocates as the smooth-running machine by which humanity has gradually become enlightened, and the only reliable source of knowledge. There are even attempts to show that science can be the source of our moral values. Privately, many of science's champions are no doubt more sophisticated, but in public they often oversimplify.

In fact, as this book will show, the borders between science and other disciplines (such as philosophy) are porous, and that is a good thing. Human understanding in general can often proceed only by the mistaken promotion of error. And, most crucially, good ideas can be found, but then rejected or ridiculed for decades or centuries or millennia before they are finally rediscovered. Yes, the past is riddled with error and fraud. In other words, it is much like the present. But it also contains surprising gems that lie waiting to be unforgotten.

Some of the ideas in this book have lately re-emerged in fields of hard scientific research – the notion that mice can inherit a fear of the smell of cherries, say, or that our universe is just one of an eternally bubbling multitude. Some, on the other hand, might just strike you as loopy – the idea that electrons have free will, or that there are infinitely many donkeys. But all of these ideas are carefully reasoned. To be sure, as the old joke goes, there is no opinion so ridiculous that some philosopher has not held it. But as we'll see in Part II, what seems ridiculous is often as important as what seems sensible.

Soon we will just be another part of history ourselves – and what then? What of the future of ideas? As advertisements for investment products are obliged to say, the past is not a reliable guide to the future. But it's the only one we've got. Ideas that are ridiculed by the mainstream or by the experts often turn out to be just ahead of their time – but that is not licence to believe any old rubbish today. And ideas that seem obviously right today will no doubt be ridiculed in years to come – but that is no reason to abandon all hope in our own judgement. Aristotle thought that slavery was acceptable because some people were slaves by nature. In Part III of this book, I'll ask: what is our equivalent conventional wisdom that will embarrass future generations on our behalf? And what apparently ridiculous ideas are we not taking seriously enough right now?

The art of rethinking, and rediscovery, lies in questioning our ideas of authority, knowledge, judgement, right and wrong, and the processes of thinking itself. Ideas cannot be pinned down like butterflies: they come from people living and thinking through time, and are passed among us down the centuries. The same idea can be bad at one time and good at another. An idea can be bad in the sense of incorrect, but nonetheless good because it is the necessary stepping stone to something better. More generally, rethinking suggests that an idea can be good in the sense of useful, even if it is bad in the sense of wrong. It can be a *placebo idea*. Outrageously, it sometimes might not matter whether an idea is true or false.

The pioneering nineteenth-century psychologist William James, brother of the exquisitely verbose novelist Henry, is often credited (perhaps apocryphally) with this ironic diagnosis of an idea's faltering progress to acceptance: 'When a thing is new, people say: "It is not true." Later, when its truth becomes obvious, they say: "It is not important." Finally, when its importance cannot be denied, they say: "Anyway, it is not new."'[14] That transition, it turns out, can take an awful long time – and cause an awful lot of sorrow along the way.

The world of ideas is a moving target. What follows is a freeze-frame. It's easy to picture ideas as static packages of thought that can be definitively judged. But that is not really accurate. Like a shark, an idea needs to keep moving to stay alive. An idea is a *process* just as much as a thing. And that process is seldom one long, linear march to enlightenment. If we are not constantly rethinking ideas, we are not really thinking. As the French say, *reculer pour mieux sauter* – if you step back first, you can jump further. The best way forward can be to go into reverse. And the best new ideas are often old ones. As we'll see next, that even goes for cutting-edge microsurgery and modern warfare.

Part I: Thesis

2

The Shock of the Old

When new circumstances require old thinking.

'If men could learn from history, what lessons it might teach us!'
– Coleridge

It's not news that wisdom from the past can still be useful today. When circumstances change, however, it seems that our thinking must change too. A new situation, it is tempting to think, demands new ideas. The invention of the tank in the early twentieth century, for example, required a change in thinking about battles. During the First World War, the early tanks were cumbersome and of limited use, but some people – such as the British army officer John 'Boney' Fuller – quickly foresaw that mechanised warfare would substantially change the conduct of future conflicts.[1] But a novel situation can also open up a new space for old ideas – and they might turn out to be the best ones for the job. Paradoxically, the best response to new circumstances may be a return to old ways of thinking.

In the early twenty-first century, for example, the US was preparing to fight a new kind of war. Rather than invading with tens of thousands of ground troops, it would infiltrate the country with small numbers of Special Forces soldiers, who would team up with rebel forces in the interior. The Green Berets, as the Special Forces are known, went in bristling with the latest lethal technologies and supported by laser-guided air strikes. But in order to do their job, they also needed to revive a warfighting tradition that had lain dormant in the US military for a century. The tank and other motorised vehicles had made cavalry charges obsolete for ever – or so it seemed. Until American soldiers rode into battle in Afghanistan on horseback.

On the hoof

The US war in Afghanistan started in October 2001, when Captain Mitch Nelson's group of twelve Special Forces commandos flew into Afghanistan on a blacked-out Chinook helicopter that contained, among other things, a dozen sacks of horse-feed.

The oats were for the Northern Alliance warlord, General Abdul Rashid Dostum, whom the Special Forces were to support against the Taliban. It turned out that Dostum actually didn't need more food for his horses, but he did appreciate the six bottles of Russian vodka the Green Berets brought. Most of all, he said, what he needed was bombs.[2] And that's how it was going to work: American bombs and Afghan cavalry against Taliban strongholds. Northern Alliance fighters galloped towards Taliban lines, firing AK-47s over their horses' heads, while the US teams radioed down precision-guided air strikes on the enemy positions that were studded with Soviet-era tanks, anti-aircraft cannons, and machine guns.

And how did the US teams themselves get into position? Why, on horseback, of course. It was still the best way to get around a lot of the country's inner terrain. 'If we wanted to move, horses were the only way,' one soldier remembered – even though most of them had never ridden before.[3] Captain Nelson recalled giving his team a quick riding lesson once they had met up with General Dostum. If they fell out of the saddle with a boot caught in a stirrup, he warned, they would have to shoot the horse to avoid being dragged to death along the rocky ground. (In the event, that wasn't necessary.) It soon became surprisingly expensive to equip American soldiers with a ride: the price of horses at the market quickly tripled to a thousand dollars each.[4]

A photo that was made public shows ten Special Forces soldiers on horseback crossing an arid slope, wrapped in headscarves against the wind and sun. One remembered how sparks flew from the horses' hooves as they scrambled over rocks. They took high mountain paths single-file, with five-hundred-foot drops to one side. One soldier's horse fell on him and broke his back – he took two shots of morphine and got right back into the saddle to complete the mission. The men became attached to their steeds. One soldier led his exhausted, faithful horse on foot the last ten miles back to base.[5] Another cavalry commander, Lt Col Max Bowers, later said it was like something out of the Old Testament. 'You expected Cecil B. DeMille to be filming and Charlton Heston to walk out.'[6]

And the horses were not only used for transport. They were used as attack vehicles too. On at least one occasion, American soldiers rode into battle alongside their allies in order to ensure that the cavalry charge succeeded.[7] According to one of Captain Nelson's declassified reports: 'I am advising a man on how best to employ light infantry and horse cavalry in the attack against Taliban [tanks], mortars, artillery, personnel carriers and machine guns – a tactic which I think became outdated with the invention of the Gatling gun [. . .] We have witnessed the horse cavalry bounding [. . .] from spur to spur to attack Taliban strongpoints – the last several kilometers under mortar, artillery, and sniper fire.'[8] A cavalry charge backed up by satellite phones and laser-guided bombs – it's a vivid example of an old idea being resurrected with a new twist. But as it went on, the Afghan campaign also illustrated all too clearly the dangers of forgetting wisdom from the past.

You've probably heard the military adage 'No plan survives contact with the enemy.' It might sound to our ears like the sort of thing a wily, cynical Marine Corps sergeant of the mid-twentieth century would say. (General Electric CEO Jack Welch cited it, and it was updated pugnaciously by Mike Tyson, who once declared: 'Everybody has a plan till they get punched in the mouth.')[9] In fact the phrase was coined in the nineteenth century by the Prussian commander and military theorist, Field Marshal Helmuth Karl Bernhard Graf von Moltke. Meanwhile, we might associate 'hearts and minds' with optimistic descriptions of recent US campaigns in the Middle East, but that phrase comes from the memoirs of Sir Gerald Templer, the British commander who led a counterinsurgency campaign in Malaya in the 1950s. As he put it: 'The shooting side of the business is only 25 percent of the trouble, and the other 75 percent lies in getting the people of this country behind us.' The answer lay, he wrote, 'in the hearts and minds of the people'.[10] Counterinsurgency strategy is a lesson that those who command modern armies seem to have to learn again and again.

How do military intellectuals make progress in military strategy? Inevitably, by looking to the past. It is sometimes cynically said that generals are always fighting the last war instead of the present one, but then military history is the only possible guide to the military future. The evolution of military doctrine is always a form of rethinking, as theorists reconsider confrontations from the past and seek to draw from them lessons relevant to the present and future of warfighting. In our time the intellectual power of military thinking is often jokily dismissed, as with the hoary line that 'military intelligence' is an oxymoron. (The surprisingly robust canard that the Iraq War came about because of an 'intelligence failure' helps to buttress such attitudes.) But the evolution of military doctrine is a careful and scholarly pursuit, the practitioners of which seek constantly to learn better what history has to teach. As Sir Lawrence Freedman, Professor of War Studies at King's College London, shows in his *Strategy: A History*, at times of crisis modern strategists have repeatedly and consciously returned to the classics of military thinking.

In this context, the idea that the past is now irrelevant can be a particularly dangerous one. During the 1990s, a school of US theoreticians proclaimed what became known as the 'Revolution in Military Affairs'. Improvements in technology such as surveillance and 'smart' weapons would, it was said, give the US unprecedented dominance in future conflicts, to such an extent that the proverbial 'fog of war' itself would dissipate and easy victory would be assured.[11] In a 2002 article in *Foreign Affairs*, Secretary of Defense Donald Rumsfeld extended the concept of this revolution to include the inspiring picture of the Green

Berets riding into battle on horseback in Afghanistan. That showed, he wrote, that 'a revolution in military affairs is about more than building new high-tech weapons – although that is certainly part of it. It is also about new ways of thinking and new ways of fighting.'[12]

It would not be long, however, before the experience of the US in Iraq as well as Afghanistan showed that his version of the RMA – what was for a while known as the 'Rumsfeld Doctrine' – was altogether too revolutionary to secure lasting satisfactory outcomes in reality. The enemy regimes were easily toppled, but Rumsfeld's insistence on doing so with far fewer forces than would previously have been considered necessary caused a problem when the liberated populations refused to stay pacified. The initially successful campaign in Afghanistan was the first time in US military history that a war had been led by Special Forces operatives, who were trained as guerrilla fighters.[13] The much bigger conventional force that invaded Iraq still numbered far fewer than observers in the State Department and elsewhere considered prudent. At length, in both Afghanistan and Iraq, the Americans found themselves fighting an insurgency with insufficient manpower. And this was all the more of a problem since no one in the US military had thought seriously about counterinsurgency strategy for more than thirty years.[14]

This neglect was itself a legacy of Vietnam – as was the decades-long sidelining of Special Forces, whose soldiers had committed some of that war's most notorious depredations. After Vietnam, the US decided it would never again get itself into such a quagmire, and instead military thinkers began to concentrate anew on the theory of great-power battles such as might (as it was then thought) break out between the US and the USSR in Europe. This took them back to the Prussian (and then Russian) general Carl von Clausewitz, the influential nineteenth-century theorist of massed forces meeting on a defined battlefield. In his great work, *On War*, Clausewitz critically analysed the campaigns of Frederick the Great and Napoleon, and characterised war as a 'dynamic interplay of politics, violence, and chance'.[15] It was he who famously defined war as the continuation of policy by other means. (You sometimes hear this as 'the continuation of politics by other means', but the politics in question is that of international power struggles rather than domestic squabbling.) The problem with Clausewitz, however, was that not only was *On War* incomplete and internally inconsistent (he died before finishing it), but no one was quite sure what one of his central concepts actually meant. This was his idea of the enemy's *Schwerpunkt*, or centre of gravity: the source of the enemy's power. At one point Clausewitz defines it as the enemy's main army; yet he also allows that there might be several such centres. As Western military thinkers of the late twentieth century returned to Clausewitz, they expanded his concept of centre of gravity – now referred to, naturally, by the military initialism COG – so that it could be a source of strength or a source of weakness, a single target or several, and a physical, psychological, or political phenomenon.[16]

While COGs ramified senselessly in military thinking through the 1980s and 1990s, a different kind of problem awaited Western powers on the ground. By 2004, the US occupation of Iraq was already in trouble, so there was a sudden and concerted turn to the literature of past insurgencies. Military thinkers quickly dusted off their copies of Gerald Templer and other manuals of guerrilla warfare. It was only in 2006 that Lt Gen. David Petraeus – later to command the 'surge' – wrote in Fort Leavenworth's house journal, *Military Review*, that counterinsurgency should concentrate above all on 'efforts to establish a political environment that helps reduce support for the insurgents and undermines the attraction of whatever ideology they may espouse'. Or, in other words, winning hearts and minds. Petraeus's guidelines may have been couched in modern jargon ('Cultural awareness is a force multiplier'), but they were old wisdom, as he showed by citing classics such as *Seven Pillars of Wisdom*, the 1922 account by T. E. Lawrence ('of Arabia') of his experiences fighting as a British officer alongside Arabs rebelling against the Ottoman Turks. (Lawrence memorably described counterinsurgency as like eating soup with a knife.)[17] Petraeus's article was entitled 'Learning Counterinsurgency', but it might justifiably have been called instead 'Relearning Counterinsurgency'.

Because it had been forgotten, the 'hearts and minds' lesson had to be laboriously rediscovered, at enormous cost. That is what happens when old ideas are consigned to irrelevance. Keep them around, and they might save lives.

The cavalry turned out to be indispensable to the military aims of a superpower in 2001. Similarly, the US was determined to fight the wars in both Afghanistan and Iraq in a revolutionary way, using high technology and fewer troops. But these new military circumstances demanded yet more old ideas: lessons from the history of counter-insurgency that, in the event, were revived all too belatedly.

Bleeding edge

It's nice to think of modern medicine, like modern war, as hi-tech, almost impersonal. Our medications are engineered on a molecular level. Keyhole surgery, computerised imaging, and robotics promise magical, almost uninvasive modes of health maintenance. So why are doctors using leeches again?

The medicinal leech has three jaws and a hundred teeth. With them it saws into the skin and injects anaesthetics to avoid disturbing its meal, chemicals that dilate the blood vessels in order to get a better flow of the good stuff, and anticoagulants to stop the blood clotting and impeding its feast. Then it starts to suck. A leech can take in up to ten times its own body weight in human blood. Nightmarish, right?

The therapeutic use of leeches has a long history. It is recorded in ancient Indian and Greek handbooks of medicine. In medieval and early-modern European medicine, patients were bled for a wide variety of supposed imbalances to the bodily humours. Do you have a red skin from fever? Too much blood: leech. Acting flighty? You're probably too sanguine, which means you have too much blood: leech. (For their remarkably regular recourse to bleeding, doctors themselves were often called leeches.) In the nineteenth century there was a Europe-wide craze for leeching, following the theory of a Napoleonic doctor who held that all illness resulted from inflammation of the intestines: starvation and bleeding, he believed, would purge the body of the toxic effects. Leeches were considered the ideal treatment for everything from nymphomania to tuberculosis. The United States imported millions of leeches from Germany, because American leeches were not as good at sucking blood.[18] But eventually chemistry and biology prevailed, and by the beginning of the twentieth century leeches had been consigned to the disgusting history of unscientific medicine.

Then, in 1985, a dog bit the right ear off a five-year-old boy from the Boston suburbs. The boy's physician, a plastic surgeon named Joseph Upton, managed to reattach the ear, but it began to turn black and die: blood could get into it but not out again, because of clotting in the veins. Blood-thinning agents didn't help; nor did lancing the ear. Luckily, however, Upton remembered an article he had once read about the therapeutic effects of leeches on congested tissue. He found the contact details of a company, Biopharm, that grew and sold leeches. (It had been set up only a couple of years earlier by a zoologist called Roy T. Sawyer, who had written a definitive three-volume study of the biology and behaviour of leeches, and suspected they were due a medical comeback because of all those interesting chemicals in their saliva.) Biopharm was in Wales, so thirty leeches were flown across the Atlantic to Boston. Joseph Upton attached two to the boy's congested ear, and in minutes it began to recover its healthy colour. After a couple of days the organ was fine, and Upton became the first doctor to have successfully reattached the ear of a child using microsurgery.[19] Plus a couple of vampiric slugs.

Upton's rediscovery had in fact been predated by others; remember, after all, that he'd seen an article on the subject. Two plastic surgeons in Yugoslavia announced the positive results of their experimentation with leeches to treat vein congestion in the *British Journal of Plastic Surgery* in 1960, though they concluded that other methods should be sought.[20] In 1972 a French surgeon, Jacques Baudet, successfully used leeches to prevent post-operative blood-clotting, and his techniques were imitated in France and the UK.[21] Baudet's work was described in the *New York Times* in 1981, and it may have been this report, if not the earlier medical articles, that Upton remembered seeing four years later when faced with a child's ear turning black.[22] In any case, third time lucky – it was Upton's sensational success that made the headlines and ushered in the real leech renaissance. Leeches became widely used in the US, and in 2004 the FDA approved leeches as a 'medical device', enabling new leech-farming companies from other countries to enter the market.[23] Today leeches are frequently used in reattachment operations, skin grafts, and reconstructive plastic surgery, since they are so good at keeping the patient's blood flowing in the damaged area, which helps the veins to knit together again.

The active anticoagulant in leech saliva, a protein named hirudin, can be produced separately, and Biopharm, which is still a going concern, has isolated and resynthesised many other active and useful compounds from the medicinal leech and other species – including the terrifying giant Amazon leech, which is the length of your forearm and stabs its prey with a six-inch needle proboscis. But the humble old medicinal leech still has a winning combination of effects, plus it is cheap – and automatically reproduces itself. Perhaps even more surprisingly, it turns out that leeches also relieve symptoms of osteoarthritis when applied to the knees, because of anti-inflammatory and other compounds in their saliva that are still not fully understood.[24] They have fewer side effects than the standard treatment of non-steroidal anti-inflammatory drugs, and they are more effective in relieving pain and stiffness than the best topically applied medication.[25] And so one traditional use of leeches, in Ayurvedic medicine and other systems – to relieve pain and inflammation – had it right all along.

As it happened, Joseph Upton had previous experience in resurrecting disgusting medical practices from the past. He had served as an army doctor posted in Augusta, Georgia during the Vietnam War, when many returning soldiers were suffering from badly infected wounds. Upton knew that Civil War doctors had used maggots – they feed only on dead flesh, and so are very effective at debriding (cleaning out) unviable tissue from wounds. He took a chance and used maggots on his soldiers – with great success, until the brass heard about it and threatened him with a court martial. Years later, when Upton had his idea to save the boy's ear, he remembered that experience, and decided not to tell his superiors that he planned to use leeches.[26] He just went ahead and did it himself. Sometimes, a rethinker has to break the rules.

The rediscovery and re-evaluation of traditional remedies is quite common in modern science. Traditional Chinese medicine, for example, is largely taken to be ineffective by Western researchers. And yet it set one woman on the path to a Nobel Prize. In 1969, a Chinese pharmacologist named Tu Youyou began working on a secret government project to search for new anti-malaria drugs. Extracts from the plant *Artemisia annua* (sweet wormwood) seemed to be good candidates, but the results were inconsistent. So Tu went back to the ancient TCM literature for clues. In a textbook from the fourth century CE, she found a herbal-recipe remedy based on *Artemisia*. It contained the information that eventually set her on the right path to a reliable extraction method.[27] The book said that wormwood should be soaked in water, but Tu had been boiling it. Now she realised that she must have been damaging the active ingredient somehow. After all, the ancient doctors could have boiled it as easily 1,400 years ago, but they didn't. So Tu switched to using an ether-based solvent, and tested the results on mice and then herself, before performing further clinical trials. It was soon clear that she had 'cured drug-resistant malaria'.[28] Tu had created a powerful new drug, artemisinin, which is now the main anti-malarial medication worldwide. For that breakthrough she was awarded the 2015 Nobel Prize in Physiology or Medicine.[29] Now that is rethinking in action. And it all happened because of a conscious decision to return to old ideas in the face of new circumstances.

The government project Tu was working on had been set up in 1967, not long after the Cultural Revolution began closing many universities and 'purging' intellectuals, including China's top malaria researcher. But in the mean time China's allies in Vietnam, the Vietcong and North

Vietnamese Army, were suffering huge casualties from malaria epidemics, which often killed more soldiers than actual fighting. Malarial resistance to the available drugs, in both Vietnam and southern China, was also increasing rapidly: the most powerful current treatment, chloroquine, was increasingly hit-and-miss. So Ho Chi Minh persuaded the Chinese government to try to develop new ones. A top-secret military research group, Project 523, was set up, with one section of it operating under a specific remit to look to the past, in the form of traditional Chinese medicine. And that was how Tu Youyou rediscovered an old idea that was right for a new era. Her work went on to save millions of lives all over the world.

The Stoical attitude

Modern people need the most up-to-the-minute, cutting-edge psychological therapies, designed for their uniquely busy and confusing lives. Freudian psychotherapy – the interrogation of feelings and personal history conducted over years of weekly or more frequent sessions – has struggled to demonstrate its clinical efficacy, and anyway who has the time? Instead the treatment of the age is Cognitive Behavioural Therapy: a set of rules and self-help techniques that can almost be reduced to an algorithm. (Indeed, it is sometimes administered by computers.) CBT trains the subject to recognise negative patterns of thinking that lead to fear and anxiety and to replace them with more realistic, practical assessments. After an unpleasant social interaction, for example, someone might habitually think: 'People are always unpleasant to me because I am an unlikeable person.' A CBT therapist would encourage recasting the initial observation neutrally (maybe that one other person was being rude because he was irritated by something else entirely), and resisting the habit of jumping to a 'totalising' conclusion of negative self-image. This style of advice has had great clinical success, to the extent that CBT is, for example, the go-to talk therapy recommended by the UK's National Institute for Clinical Excellence. It is no-nonsense, 'evidence-based' treatment: quintessentially modern. Yet at the same time, it has very deep roots in the pre-scientific age. And its original inspiration is now making a comeback on its own terms.

You've probably heard the phrase 'Every day, in every way, I'm getting better and better', which might sound like the height of over-optimistic positive thinking. (In *every* way? Really?) In fact this mantra originates with a fascinating character called Émile Coué, a French pharmacist who became one of the first global stars of self-help literature. He hit the big-time in America in the 1920s with his book *Self-mastery through Conscious Autosuggestion*. Repeating such positive phrases, he said, provokes a positive 'autosuggestion' – a kind of productive subconscious attitude – that will improve your mental and physical health.

If that sounds flaky, consider the following. 'Suppose our brain is a plank in which are driven nails which represent the ideas, habits, and instincts, which determine our actions,' Coué suggests. 'If we find that there exists in a subject a bad idea, a bad habit, a bad instinct – as it were, a bad nail, we take another which is the good idea, habit, or instinct, place it on top of the bad one and give a tap with a hammer – in other words we make a suggestion. The new nail will be driven in perhaps a fraction of an inch, while the old one will come out to the same extent. At each fresh blow with the hammer, that is to say at each fresh suggestion, the one will be driven in a fraction further and the other will be driven out the same amount, until, after a certain number of blows, the old nail will come out completely and be replaced by the new one. When this substitution has been made, the individual obeys it.'[30]

This is recognisably the same idea as that behind modern CBT. Both systems train patients to recognise automatic negative thoughts and consciously replace them with more reasonable reactions, so that the bad thoughts are finally driven out. It was also the therapeutic principle of Coué's contemporary Paul Dubois, a neuropathologist who founded a 'rational persuasion' school of psychotherapy, and used it on none other than Marcel Proust. And that principle – that focused reason ('cognition') can be used to overcome emotional disturbance – was already more than two millennia old. It was the principle of the Stoic philosophers. And now they're back too.

We use the term 'stoical' in everyday conversation to mean something like stiff-upper-lipped, controlled, suffering without complaining, or even emotionless, like Mr Spock. But the ancient Stoics were more cheerful than this – or at least they tried to be. The movement was founded in Athens by Zeno of Citium, and its literary luminaries subsequently included Epictetus and the Roman emperor Marcus Aurelius. Its central insight is often thought to have been best expressed by Epictetus: 'Men are disturbed not by things but by the views which they take of them.' In other words: you can't change what has just happened to you, but you can change what you think about it. And hence, what you feel about it.

Two thousand years later, Albert Ellis, the founder of Rational Emotive Behaviour Therapy and considered one of the co-pioneers of modern CBT along with Aaron Beck, wrote in 1962 that a person is not affected by things and events themselves but 'by his perceptions, attitudes, or internalized sentences *about* outside things and events'. (An 'internalized sentence' here would be the same kind of thing as Émile Coué's unconscious autosuggestion – a bad nail to be driven out by a good one. In fact, Albert Ellis had himself read Coué.)[31] 'This principle,' Ellis acknowledged, 'was originally discovered and stated by the ancient Stoic philosophers.' Aaron Beck, too, explicitly acknowledged the historical roots of his programme. 'The philosophical origins of cognitive therapy can be traced back to the Stoic philosophers,' he wrote. 'Control of most intense feelings may be achieved by changing one's ideas.'

Perhaps the most surprising effect of the modern popularity of CBT has been a revival of ancient Stoicism itself. Every year now there is a 'Stoic Week' organised by the University of Exeter, with a week of talks and workshops in London, and the participation of thousands of people all over the world who read the Stoics, practise exercises, and take part in a survey. People who took part in 2014, wrote the biologist, philosopher, and neo-Stoic Massimo Pigliucci, reported 'a 9 percent increase in positive emotions, an 11 percent decrease in negative emotions and a 14 percent improvement in life satisfaction after one week of practice (they also did longer term follow-ups, which confirmed the initial results for people who kept practising)'.[32]

Stoicism as originally formulated is not precisely the same sort of thing as CBT. The general principles and some specific techniques are remarkably similar, but Stoicism also encompasses an entire logical and metaphysical system, as well as some meditative tools that have not been adopted in modern therapy – although some practitioners think they should be. Stoical texts, writes Donald Robertson, the therapist and author of *The Philosophy of Cognitive-Behavioural Therapy*, 'contain many specific psychological techniques or exercises, most of which are consistent with modern CBT, and some of which have been forgotten or neglected by modern psychotherapists, though still relevant today'.[33] One that he laments our losing touch with, for example, is the systematic meditation on death, or (in Greek) *melete thanatou*.[34] Some modern research, indeed, suggests that 'the awareness of mortality can motivate people to enhance their physical health and prioritize growth-oriented goals; live up to positive standards and beliefs; build supportive relationships and encourage the development of peaceful, charitable communities; and foster open-minded and growth-oriented behaviors'.[35] Another technique that Robertson has revived in his own practice is encouraging patients to adopt a panoramic aerial view of things, dissociating from one's personal concerns by intensely 'visualizing the world as if seen from very high above'.[36]

It may be useful, alternatively, to adopt the excellent example of Marcus Aurelius and say to oneself every morning: 'I shall meet today ungrateful, violent, treacherous, envious, uncharitable men [. . .] I can neither be harmed by any of them, for no man will involve me in wrong, nor can I be angry with my kinsman or hate him; for we have come into the world to work together.' Alternatively, you can try Hierocles's circle, which involves imagining yourself in an ever-widening circle of friendly concern: starting with your family and friends, then neighbours, all the inhabitants of your city, all your compatriots, then every human being on Earth, and then all of Nature itself.

Most difficult, perhaps, is the *premeditatio malorum*, which means meditating on very bad things that might happen to you and trying to adopt a serene perspective on them. You attempt to vividly imagine suffering a serious physical injury or emotional setback, for example, and then try to view it as a 'dispreferred indifferent'. This means that it would be better if it didn't happen, but if it did it would still be 'indifferent' in the sense that it would not harm your moral worth or your personal integrity. This, Pigliucci points out, 'is very similar to an analogous practice in CBT meant to allay one's fears of particular objects or events'.

Lastly, you can follow Seneca and pose the following tough questions to yourself in the evening: 'What bad habit have you put right today? Which fault did you take a stand against? In what respect are you better?'

As we can see, Stoicism is not mere endurance; nor is it warm and fuzzy. (Indeed, Friedrich Nietzsche called it 'self-tyranny'.)[37] But, despite the existence of aeroplanes and smartphones, it's no less applicable to the challenges of daily life than it was two and a half thousand years ago. You might not want to subscribe to all the Stoics' beliefs: they argued, for example, that the universe was governed by the principle of Logos, the outpouring of a divine creative mind – though some modern physicists, as we'll see, are not troubled by such a concept. But the Stoics as a whole were an agreeably progressive bunch. They argued for sympathy (though not liberty) for slaves, and were splendidly cosmopolitan. As Epictetus put it, everyone is a citizen of two worlds: not just the 'small city', which is his immediate political community, but also the 'great city', which is the entire universe.[38]

The Stoic philosophy is a great engine of reason. It encourages us to take a different view – literally, an aerial view – of our place in the great scheme of things, and it reminds us of the control cognition can exert over emotion: the power of thinking to improve itself.

Circumstances and technologies change, but two thousand years is the merest blink of an eye in evolutionary terms: the human mind has probably not changed all that much. So ancient traditions can be resurrected for ultra-modern people: enter a smartphone app called the PocketStoic. Or they may be shinily repackaged, as is the case with the popularity of 'mindfulness', an approach drawn from Buddhist or Taoist meditative traditions which is now sold as a way to improve our productivity and happiness in the uniquely pressurised modern workplace. And what is true of the continuing relevance of old therapies to the modern age is also more true than seems immediately obvious in other fields, such as warfare and medicine. Against the assumption that modern times require entirely novel thinking are ranged the Special Forces on horseback, the leeches draining that boy's ear, and the modern rediscovery of powerful self-help techniques invented millennia ago.

3

The Missing Piece

How an old idea can become relevant again with the discovery of a new piece of the puzzle.

'To achieve anything really worthwhile in research, it is necessary to go against the opinions of one's fellows.' – Fred Hoyle[1]

Tesla Motors was able to revive the electric car in part because of advances in battery technology. The short distance that early electric cars could travel before needing to be recharged limited their appeal when compared to the far superior autonomy of gasoline-powered vehicles. But by the twenty-first century, lithium-ion technology had made batteries unimaginably better than those of a hundred years ago. Early Tesla prototypes were actually built using hundreds of ordinary phone batteries connected together, and Musk's company refined the idea to build its own battery packs that enabled its cars to go hundreds of miles between charges. The perennial problem of the electric car, 'range anxiety', was banished.

Modern battery tech was the missing piece that made the idea of the electric car practical once again. A similar dynamic works for other ideas, which must be upgraded if they are to become newly viable. A new piece of the puzzle, or the revelation of a hidden mechanism, can bring an idea that was once deeply unfashionable or positively ridiculed back to the cutting edge of theory – in modern science and in the Royal Game.

The Berlin Wall

It's a dark, wintry afternoon in London. Behind the bland corporate façade of the Olympia Conference Centre, ten of the world's best brains are battling it out for a week at the London Chess Classic tournament. 'The grandmasters are playing in the auditorium, that way,' the woman checking passes indicates in a respectful hush. There is also a bigger tournament taking place in a hall to the side, which is open to everyone. As I check my coat I hear a man behind me say: 'John, are you playing what we discussed last night?' His friend, an American, replies: 'Uh, most likely – I'm not sure I'm allowed to talk about the game.' The first guy laughs and says: 'That's why I was trying to be obscure!'

In the commentary room, everyone is talking about the games – the grandmaster games, that is, and well out of earshot of the combatants themselves. Daniel King, a stylish Englishman with steely Byronic locks, and Jan Gustafsson, a sardonic, fresh-faced German, are seated behind a white table surveying the early moves of each of the five games in progress, running through possible developments and giving opinions. (As grandmasters themselves, they are well qualified to explain what is going on, although sometimes even they confess bafflement at the refinements of 'super-grandmaster' play at the very top level.) Two large video screens show the boards and players, while the commentary is also being broadcast live online to a global audience. King and Gustafsson, tag-teaming with another pair of commentators, will be calling the shots until all the games are finished – which could be seven hours from now.

Inside the auditorium itself, the atmosphere is thick with mental power. On the stage, the players sit at five tables, while the positions of the games are projected on to a big screen at the back. The saturnine Russian player Alexander Grischuk has left his own game, against the former world champion Vishy Anand, to stand loomingly over another. The US champion Hikaru Nakamura, facing a tricky situation, is curled right over the board, head in hands, ankles intertwined. His opponent, the snappily dressed Armenian Levon Aronian, is ambling around looking at other boards too. At one point Magnus Carlsen, the Norwegian wunderkind and current world champion, strolls over to take a look at Nakamura's game, standing right behind the empty chair of his opponent. He thinks for a few seconds, smiles to himself, and then wanders off.

Back in the commentary room, when there is nothing new on the boards, King and Gustafsson are talking about personalities and psychologies. What did one player mean when he described another player's style as 'bleak'? This other player is not likely to offer an early draw, because he's just eaten a protein bar. But what might really confuse anyone wandering in who is not a chess nerd is this: there are also a lot of in-jokes about Berlin. A few days previously some wag had christened the event the 'Berlin Chess Classic'. The commentators are pleased that there is only one Berlin today. 'Yeah, the players had a meeting on the rest day,' Gustafsson says drily. 'They decided: no Berlins any more.' It turns out to be an optimistic joke: there are many more Berlins to come. So why is everyone in London so down on Berlin?

Success in top-level chess depends on two kinds of accomplishment. There is the necessity to work things out in real time, facing your opponent, calculating further, outplaying the adversary – that is 'over-the-board' play. Just as important is what players call 'preparation'. This involves the pre-match research into opening strategies. As we all know, White has the first move. Depending on what he plays, Black then has several viable options for the second move. Depending on what Black plays, White then has many options for the third. And so on. Each combination of the first handful of moves defines an 'opening', with a name such as the Queen's Gambit or the Sicilian: there are many variations of each of these too. Once you are only a dozen moves into the game, there are millions upon millions of possible positions that can arise from following the rules of chess – what are called *legal* positions. But are they any good? Would you actually want to play on from such a position? The accumulated chess wisdom as to which of these variations are in fact playable – not giving a big advantage to the opponent – is called 'theory'. And theory is terrifically important.

Some complain that modern professional chess is all about memorisation. That's not true. Sometimes, yes, a player can win a game in twenty-odd moves without getting out of his preparation – without having to think over the board – because his opponent has lost his way in some very complicated theory. On the other hand, chess is so mind-bendingly complicated and rich in possibilities that even with the state of modern chess theory, built on databases of millions of games, it is still perfectly possible, within the first ten moves of the game, to play a viable move that has never been seen before. And then the real-time struggle begins.

So chess involves two battles. The over-the-board battle is still primary. But the pre-match battle of preparation can give a determined player an important advantage. This is particularly true in match play – when two players contest a series of games against one another – as opposed to tournament play, when a player accumulates points through games against many others. Match play is head-to-head and gives you repeated chances to use your preparation against the same opponent. That is how the World Championship is decided. And it is one World Championship match in particular – the one that took place in London in 2000 – that explains why commentators in the same city fifteen years later were still making jokes about Berlin.

It was a duel of the imperial, world-beating master against his one-time protégé and assistant. The master was the Russian genius Garry Kasparov. These days, he mainly makes the news as a pro-democracy, anti-Putin commentator on Russian politics. Back in the year 2000, he was the world's greatest chess player, and widely considered one of the very best in the history of the game. At thirty-seven, Kasparov had been world champion for fifteen years, fending off all comers in brutally swashbuckling style. In match play he was invincible. And the twenty-five-year-old Russian challenger, Vladimir Kramnik, needed to find a way to beat him. Kramnik, the son of a sculptor and a music teacher, had studied under Kasparov as one of a group of talented juniors, and in 1995 he had served as one of Kasparov's seconds, or assistants, in the champion's previous successful defence of his title. Now the two men were to play sixteen games to decide the world championship, with a prize fund of two million dollars. It happened in October, in London. Eleven years after the fall of communism. And yet it became known as the match of the Berlin Wall.

Few thought Kramnik had much of a chance. But he had been inspired by the story of another sporting underdog two years earlier. He was going to use a lesson from ice hockey. In the 1998 Winter Olympics in Japan, the USA, Canada, and Sweden were the most hotly tipped squads. The Czech Republic team, meanwhile, was on no one's list of the tournament favourites. They had one decent forward but no other really world-class players – except their goalkeeper, Dominik Hašek. So what is your strategy when your best player is the stopper? Hunker down in defence, let them throw everything they've got at you, and score on the counter-attack. And that's just what the Czech Republic did. In the quarter-finals they beat the USA 4–1. As the chastened American coach, Ron Wilson, said: 'Sometimes in our sport, one individual can make a big difference. A good goaltender can knock a team out.'[2] It happened again in the semi-final against Canada, deadlocked at 1–1 after extra time. There was a penalty shootout, and Dominik Hašek stopped all five shots by the Canadians. In Prague, fans held up banners reading 'Hašek for President'.[3] And it happened once again – perhaps most impressively – in the final against Russia. The Russian forward Pavel Bure had scored five goals single-handedly in their semi-final demolition of Finland. But in the final the Czech goalkeeper shut them all down. The Czechs won 1–0. Dominik Hašek was the Dominator. No one got past him.

So, the young challenger Kramnik wondered, how do you do that in chess? If you're facing a player whose speciality is the terrifying attack, you need an impregnable defence: the chess equivalent of Dominator Hašek. And that was the critical idea (among several) that Kramnik developed during his preparation for the match: searching for defences he might employ with the black pieces that would blunt Kasparov's offensive force. With the help of his team, Kramnik looked again at an old, relatively obscure sub-variation of the Ruy Lopez (or Spanish) opening. Here, the Black player was able to encourage an early exchange of queens, so avoiding any complex all-out attack by White. Instead the variation went early into an endgame. But everyone had long thought this endgame was worse for Black, who would struggle to make a draw and could well lose. For that reason this opening had barely been used in top-level play for decades. It was considered an evolutionary dead-end. But Kramnik took a different view.

This approach, Kramnik knew, would sidestep any grand attacking plans by White. It was also very unlikely that Kasparov would have foreseen this choice and prepared his own reply. So even if Kramnik did have to play a slightly worse endgame, he knew the outcome would depend on a real battle of wits. 'I accepted that the endgame was better for White,' Kramnik said later, 'but he has to win over the board, not with his legendary home preparation – that's crucial!'[4] (It later turned out that, during his preparation for the match, Kasparov had ignored warnings from his assistant, Yury Dokhoian, that Kramnik might try to direct games into 'anodyne positions, with few pieces on the board'.)[5] Most importantly, because this particular variation of the Ruy Lopez was moribund and hadn't been played seriously for so long, the old analysis about which possible moves were better than others was out of date and could be improved with modern knowledge. So Kramnik and his team prepared new moves and ideas to plug into this old system. And in the very first game, Kramnik was able to employ this line. A surprised and confused Kasparov eventually agreed to a draw. What was the name of this opening? Why, the Berlin Defence. The name records the place of its first use – a tournament in Berlin, in 1840. But after this match it would become famous once again, as the impregnable Berlin Wall.

As it happens, the world champion and his challenger also employed different approaches to in-game nutrition. Garry Kasparov's team made sure there was mineral water and chocolate at the table; but Kramnik had what one of the match officials later called 'a huge variety of beverages and snacks, scientifically chosen by his trainer'.[6] Kramnik had also gone through an intense fitness regime involving swimming, weight training, and volleyball. ('It made a big difference to my match stamina,' Kramnik said.) But it was the Berlin that really made the difference. As the match proceeded, Kasparov threw himself again and again at the Berlin Wall, and it never crumbled. Kramnik did not lose a single game as Black, and with White he won two beautiful games against his increasingly demoralised opponent. And so Vladimir Kramnik became the new chess champion of the world.

His secret weapon had been an old idea that had long been thought obsolete, now re-evaluated and upgraded. 'With the Berlin,' Kramnik explained later, 'I was able to set up a fortress that he could come near but not breach. When others play against Kasparov they want to keep him distant. I let him in close but I knew where the limit was [. . .] Someone even compared it to Ali's "rope-a-dope" trick against George Foreman – this was a very good analogy! With the Berlin, I was able to allow him to get near, but not quite near enough, and I knew where to draw the line with the fortresses I had set up. At some point he seemed to lose all confidence trying to break down the Berlin Wall. He was still fighting as only Kasparov can, but I could see it in his eyes that he knew he wasn't going to win one of these games [. . .] This is what broke him down psychologically.'

The Berlin had barely been played for decades before Vladimir Kramnik rethought it. Afterwards it was all the rage. Much later, Kasparov himself called the improved variation an 'ingenious invention' by his opponent: 'It is only the smallest exaggeration to say that the Berlin cost me my World Championship title against Kramnik in 2000.'[7] And fifteen years on, at the London Chess Classic, the top players in the world were still using it. It had now split into a wealth of modern sub-variations, like the branches of a tree – some super-solid, some much sharper and more unbalanced. In chess, opening variations rise and fall in popularity, but this is not a matter of mere fashion. Instead, the best players work on a consensus as to what constitutes the cutting edge of chess theory – fascinated by the possibilities of a particular line, they act as a kind of competitive research group, trying to explore all the ramifications with both colours.

And the rise of a newly popular opening always comes as a result of Kramnik-style rethinking of an old, discarded idea. You look at a position that theory says is bad. You plug in a new move, a new idea – and you improve theory. In chess, innovation is inseparable from reaching into the past. Some forgotten line from the textbooks will be looked at anew and reanimated with a 'novelty' on the tenth or sixteenth move; a grandmaster will play his new discovery in an important game and win. And then everyone will pile in to try to see what the truth really is. (A grandmaster doing such research will in the modern age also be using chess software running on a powerful computer; but it is important to know when to trust the computer's opinions and when to ignore them.) During one Berlin-heavy round of the 2015 London Chess Classic, the British grandmaster Jon Speelman suggested a possible way to play for Black in one game and then said: 'Well, I don't understand the Berlin. Who does understand the Berlin?' The audience laughed. Jan Gustafsson replied, wryly: 'No one does. No one does with White either.'

And so they keep playing it, until they do.

A new plan in an obsolete chess opening can turn it into a devastatingly up-to-date weapon. And, as we have seen, the cavalry charge is still a valid military tactic in the twenty-first century – provided you have artillery and laser-guided bombs to back it up with. The belated discovery of a missing piece, buried deep within the molecular action of living bodies, is also the key to the revival of a scientific idea that was ridiculed for a century.

The French father of evolution

If, one crisp spring afternoon in Paris, you creep by the back entrance into the Jardin des Plantes, the grounds of the National Museum of

Natural History, you will meet a succession of French scientists of the eighteenth and nineteenth centuries, immortalised in bronze. Here is Michel Eugène Chevreul, chemist and inventor of an early form of soap, gesturing kindly towards what is now a car park, as though saying: 'By all means, station your vehicle here.' Round a curving path is the zoologist Bernardin de Saint-Pierre, having a Rodin-style think in a green glade. Seated in front of the Great Hall of Evolution is the encyclopaedic naturalist the Comte de Buffon, manhandling a pigeon with a faint Bond-villain smile. And as you wind through the gardens you will eventually meet the star of the show, presiding over the front entrance: the botanist and zoologist Jean-Baptiste Lamarck, pensive and far-seeing on his plinth.

'To the founder of the doctrine of evolution,' says the inscription in French on the front of the statue, a puzzle for any visitors who thought a certain Charles Darwin was the man most deserving of that title. Round the back of the statue, however, is a more melancholy tribute. A bronze relief shows a young woman laying a comforting hand on an old and sad Lamarck, who is slumped on a chair. 'Posterity will admire you,' the woman is assuring him. 'Posterity will avenge you, father.' But what need does one of the greatest scientists of the Enlightenment have for vengeance?

Lamarck was born in 1744, and after stints in the army, and as a bank clerk in Paris, he became an amateur botanist. In 1778 he published a book on French flora, on the strength of which he was appointed assistant botanist at the Natural History Museum. Just after the French Revolution, he had the politically astute idea to rename the Jardin du Roi (the King's Garden): so, the Jardin des Plantes it became. Lamarck was promoted from assistant botanist to professor of the natural history of insects and worms, which he knew nothing about. Nevertheless, he now had time to think deeply. It was Lamarck who coined the word 'biology'. He was also the first to propose a coherent theory of evolution.[8]

Living matter, he said, tends to organise itself in increasingly complex ways – as may be confirmed by comparing an amoeba to a dog – and species gradually adapt to their particular environments, acquiring or discarding the appropriate characteristics. (So, for example, polar bears developed white fur for camouflage in the Arctic environment.) This idea was extremely controversial and largely rejected in Lamarck's lifetime, and not only because, as one of the first purely materialist accounts of life's development, it left no room (or at least necessity) for God. It scandalised, for example, the eminent paleontologist Georges Cuvier, who was the first to identify certain winged fossils as ancient flying lizards, for which he coined the name pterodactyl, and who thought there must have been some catastrophe (analogous to the biblical flood) that killed very many creatures a long time ago. (Two hundred years later, it turned out that there had been: the asteroid that wiped out the dinosaurs.) Yet Cuvier brutally mocked Lamarck's notion that animals could transform themselves, and stoutly defended the common-sense picture of species as fixed.[9]

Lamarck's books on evolution were never successful, and he wasn't accorded the professional respect enjoyed by Cuvier and others. He eventually turned blind and died in Paris in 1829. His family was so poor at his death that his books were sold off and his body buried without a coffin, in a lime pit whose contents were regularly exhumed and transferred to the Paris catacombs. Thirty years later, Charles Darwin's *On the Origin of Species* provided the missing key mechanism – natural selection – that completed Lamarck's idea. Posterity avenged him, as his daughter had predicted it would, even if his name was now eclipsed by Darwin's.

Yet Lamarck's reputation was still not secure: indeed, in another few decades it was to plunge into the blackest abyss. For the whole of the twentieth century, Lamarck's name became a kind of ominous joke, a byword for biological theories that were not just mistaken but ridiculous. For genetically informed biology, 'Lamarckism' was everywhere denounced as unthinkable. What, then, was so absurd in his version of evolution?

How did the giraffe get its long neck? It was this question, childlike in its simplicity, that led Jean-Baptiste Lamarck to his amazing idea. The usual story one hears about his theory goes like this. Once upon a time, Lamarck reasoned, giraffes all had short necks. They happily ate the low-hanging leaves from trees. But once all the low-hanging leaves around a particular tree had been munched, a giraffe would look longingly at the leaves that were just out of reach. It would wish it had a slightly longer neck. It would somehow *aspire* to have a longer neck. And so, out of desire, it would regularly stretch its neck as much as possible to get to just one more higher-up leaf. Years of neck-stretching by the giraffe would result in a neck that was permanently just a little bit longer than it had been to begin with. And when it came time for it to have baby giraffes, the offspring would in turn inherit that slightly longer neck. And so, over a very long span of time and innumerable generations, giraffes came to have very long necks indeed.

This is an ingenious explanation for why giraffes have long necks. Unfortunately, it is wrong. This could have been appreciated as early as 1865, when the Augustinian monk Gregor Mendel presented his own revolutionary work to the Brünn Natural History Society. Having conducted careful experiments in the cross-breeding of peas, Mendel demonstrated that some 'invisible factors' determined the appearance of visible traits – such as flower colour, plant height, and seed shape – in mathematically predictable ways. It was Mendel who coined the terms 'recessive' and 'dominant' for certain traits, and his invisible factors were later christened 'genes' in 1909 by the Danish botanist Wilhelm Johannsen. (The Greek *genos* means 'birth'.) But the published proceedings of the Brünn Natural History Society were not, it appears, on everyone's must-have subscription list. Mendel's work languished in obscurity for thirty-four years. Darwin never read it. As the physicist Erwin Schrödinger commented in 1944: 'Nobody seems to have been particularly interested in the abbot's hobby, and nobody, certainly, had the faintest idea that his discovery would in the twentieth century become the lodestar of an entirely new branch of science, easily the most interesting of our days.'[10] At last, Mendel's results (and his pioneering paper) were rediscovered independently by other researchers at the turn of the twentieth century.[11] Rapidly, biologists decided that evolution couldn't work in Lamarck's way at all. Mendel's work showed that genes controlled traits, and genes seemed to be fixed at birth. So nothing that happened in an animal's lifetime could affect what it passed on to its offspring. It would simply pass on the DNA it was born with.

The accepted story became this: random mutations in the genetic code gave some giraffes slightly longer necks than others. These longer-necked giraffes would leave more offspring than the giraffes without this beneficial mutation, which could not reach as much food and so were more likely to starve before reproducing or to have sickly and malnourished babies. Rinse and repeat, in a long brutal cycle of chance and death. That was how evolution worked. Genes were destiny. Lamarck's idea that an animal's life experience could alter its reproductive legacy – the inheritance of acquired characteristics – was now denounced everywhere as a gross error made by a man who was ignorant of the real mechanism of inheritance, the gene.

Charles Darwin himself had always assumed that the inheritance of acquired characteristics must be possible, however, and some thought there had to be something to Lamarck's idea even once genetics had become the big story. Sigmund Freud, for one, consistently defended it. Freud was trained as a biologist at the peak of Lamarckian speculation in evolutionary theory during the late nineteenth century. But then the new genetics came along and Lamarckism was largely abandoned. Yet as late as 1939, in his last published work, *Moses and Monotheism*, Freud observed that 'the present attitude of biological science' rejected 'the idea of acquired qualities being transmitted to descendants'. Freud, however, could not himself agree. 'I admit, in all modesty,' he wrote, 'that in spite of this I cannot picture biological development proceeding without taking this factor into account.'[12]

By the 1920s, Lamarckism was already a heresy in mainstream biology. And things got even worse once it became associated with the pseudoscience of Stalin's favourite agricultural researcher, Trofim Lysenko. Apparently sincerely, Lysenko rejected Mendel's genetics on ideological grounds. (As the modern biologist John Maynard Smith explains, the idea of 'a gene that influences development, but is itself unaffected' is troubling to Marxism because it is 'undialectical'.)[13] Lysenko decided that genes did not exist and neither did natural selection, but that acquired characteristics were easily heritable. He claimed that rye could transform into wheat and then into barley, and that weeds would evolve into tasty food crops. In an enormous purge, more than three thousand Soviet biologists were imprisoned, sacked, or executed over the next few decades. Meanwhile, Lysenko's 'great agricultural revolution' resulted in lower crop yields. So for the rest of the century, Lamarckism came to be viewed as not just wrong but dangerous. It was a doctrine that, along with Stalin's policies, had helped starve millions to death. Around the back of his statue in Paris, Lamarck sat blind and despondent on his chair, wondering if his daughter had been wrong after all.

Cut to 2003, and a young French scientist named Isabelle Mansuy. She is working at the University of Zurich's Brain Research Institute, trying to create a model of borderline personality disorder in mice. She sets up what might seem an excessively sadistic series of experiments. At unpredictable intervals, she tosses a group of female mice, mothers of young pups, into ice-cold water (they hate water), or separates them from their young for hours, or dangles them by their tails over a long drop. In this way, Mansuy subjects the mice to chronic stress and creates some very depressed mothers. (A laboratory mouse is considered depressed if it exhibits fatalistic behaviour, such as not struggling to escape from a swimming test, or if it has lost the ability to feel pleasure, as suggested by a lack of any preference for sugared water over normal water.)[14] As the young mouse pups of Mansuy's stressed mothers mature, it becomes apparent that they are depressed as well. So far, so understandable: a troubled childhood does neither mice nor men any favours. But the kicker comes when the depressed male pups, once grown up (and still depressed), are mated with a happy group of adult females. The offspring of these unions, despite being given no contact with their troubled fathers, are also depressed little mice from birth.

Wait. That was really not supposed to happen. Depression in the adult males was an acquired characteristic, resulting from experience in their childhood. If their own children were then born depressed, that would mean an acquired characteristic had been inherited. And that would be the heresy of Lamarckism.

But that is exactly what happened. The mice had inherited stress from their parents through epigenetic means. The field of epigenetics (which comes from the Greek for 'around the gene') studies how, in response to environmental cues, chemical reactions in an animal's body actually switch genes on or off in the DNA. In particular, chronic stress, Mansuy and others have discovered, silences certain gene expressions in the brain through a process known as methylation, leading to long-term depression. And this is passed on through the germ cells (sperm and eggs) to the unfortunate offspring, in mice – and perhaps in men too. A kind of Lamarckism, then, is indeed possible.

Isabelle Mansuy later said that, while conducting her experiments – which were supported by very similar results with rats elsewhere – she had read manuscripts of Lamarck, with something of a transgressive thrill. 'The feeling I had in reading those old monographs,' she said, 'was that he was so right. He was right.'[15]

Lamarck's theory wasn't wrong in principle after all. But how exactly could the inheritance of acquired characteristics take place? No one knew. Then genetics came along: it seemed as though inheritance was now thoroughly understood, and it could not contain any Lamarckian element. But something remained to be discovered: the theory of epigenetics. That was the missing piece that made it work.

Professor Mansuy, as she is now, appears smiling on Skype from her Zurich apartment one afternoon, with birds tweeting in the background. She explains that Lamarck was even more right than she first thought. It turns out that he never actually said that giraffes got long necks because of some deliberate striving. That misunderstanding arose because of a bad translation. 'I reread the original,' Mansuy says, 'and he does not talk about a *desire* or *wish* from the giraffe, but it's just something that happened – it was adaptation to the external environment. There was a big misunderstanding about the "conscious desire" of the giraffe to grow a longer neck. What Lamarck was suggesting was that it's just the *consequence* of habituation to this environment.'

But that sounds very similar to Darwin's later theory of natural selection, right?

'Yeah, absolutely!'

'I do think,' she continues, 'like many others nowadays, that Lamarck had more developed ideas, and a lot sooner – almost fifty years sooner – than Darwin. He was unlucky, I don't know.' She laughs sadly. 'I've sometimes felt a little bit offended when people say to me: "Oh yeah, your business is just neo-Lamarckism, you're just reviving the ideas of this strange and silly Frenchman." When you really read his work he was totally revolutionary, and very courageous in his ideas.'

Revolutionary things might also come of the research Mansuy now directs on neuroepigenetics in her own lab. ('You know, doing experiments in the lab is like cooking,' she says happily. 'You need to be very creative but precise at the same time.') Because if what happens in mice also happens in humans, the consequences for our thought are enormous – in medicine and in morality.

But how exactly do we establish that inheritance of stress and depression can happen in humans? We obviously can't do the same experiments on human babies that we do with mouse pups. So we look for what scientists call a 'natural experiment': one where things have been set up by chance and circumstance to enable useful comparisons to be made. There are, unfortunately, groups of people exposed to traumatic stress in early life, because they live in countries suffering through wars, or because they are born into violent families or otherwise abused when young. Mansuy's team has been collaborating, for instance, with doctors working with traumatised soldiers, and a group of people who survived the genocide in Rwanda. From such people, researchers have collected blood, saliva, and often sperm, in order to conduct analyses. It is early days, Mansuy says, but 'I have great hope that what we have observed in mice we'll be able to observe in humans.' Indeed, 'we started by developing a mouse model purely based on human literature. We developed a paradigm which was the most closely related to traumatic stress in early life in humans. So we are just going back to where we started almost fifteen years ago.'

If stress can be inherited in humans, isn't it even worse to be unkind than we had already thought? 'Absolutely,' Mansuy replies. Because this mechanism multiplies the consequences of harmful behaviour. If you cause a person to become stressed or depressed, you harm not only an individual, but potentially their offspring. Other epigenetic research has found that descendants of Holocaust survivors have different stress-hormone profiles from those without such trauma in the family, which makes them more vulnerable to anxiety disorders. And in 2015 a study of Jewish families, some with members who suffered such trauma during the Second World War and some with no direct connection to the Holocaust, claimed to have identified specific epigenetic tags on a particular stress-related gene in both the Holocaust survivors *and* their children. Rachel Yehuda, the lead investigator, said: 'To our knowledge, this provides the first demonstration of transmission of pre-conception stress effects resulting in epigenetic changes in both the exposed parents and their offspring in humans.'[16]

But think of all the trauma that humankind has experienced since the beginning of history. Why aren't we *all* depressed, all the time? It turns out there is a silver lining. Epigenetic changes can be reversible, Mansuy reports. If you are traumatised in early life but later live with a loving family or in an otherwise positive environment, you can shrug off molecular fate. Mansuy's lab tested this idea with second-generation offspring of traumatised male mice, which were put into an 'enriched environment': cages with toys, a running wheel, and social groups. A luxury mouse hotel. And it did the job: 'We could see a reversal of symptoms after environmental enrichment,' Mansuy reports with satisfaction. 'The males whose fathers were traumatised and who had been depressed and antisocial were, after environmental enrichment, no longer able to pass on their bad phenotypes' – or individual traits, in this case depression – 'to their own offspring.' It is a sort of expiation of original sin in the mouse. And so, it seems, there may be a biological basis for hope.

The 'black box' is a concept from engineering, which describes a device whose inner workings are opaque. You can see what goes in and what comes out, but you can't observe the internal mechanism. And some ideas can be black boxes too. Lamarck's original idea that acquired characteristics could be inherited was a black box: he didn't know about epigenetics, so he couldn't be sure how his idea worked at the causal level. Once genes had been discovered, the lack of any possible explanation for Lamarckism seemed enough of a reason to reject it completely. Indeed, the German biologist August Weismann argued, in his influential 1904 work *The Evolution Theory*, that since scientists were not only unable to show that any particular acquired characteristics were inherited, but also unable to form 'clear conceptions' of any 'hypothetical process' that might enable this to happen, they were justified in rejecting the hypothesis.[17] In the event it took two hundred years to begin to identify the correct mechanisms behind Lamarck's idea. Now it seems very likely that trauma can be stamped into the epigenome. But how *exactly* does it happen? That, Mansuy says, is another 'mechanism that we don't know yet', and the focus of further research. And so, centuries on, we have finally opened Lamarck's black box – only to find that there is another black box inside it.

Maybe one day, I suggest to Isabelle Mansuy, smart new drugs building on her neuroepigenetic research will be able to address the chemical pathways associated with stress and depression? Her response is surprising. 'But perhaps that's what psychotherapy already does, right? We don't know! There has not been any longitudinal, big study about this. But I do think that psychotherapy can probably alter the epigenome.'

Currently, indeed, psychotherapy – which we are using here to mean any kind of talking therapy, including CBT – might be the most precise tool we have to treat mental suffering. Mansuy explains that, as things stand, 'Very often, drugs work by activating an alternative pathway that does not necessarily *fix* what's broken in the brain, but which compensates for it. I tend to think that psychotherapy would be more likely to address the real problem. Because if someone is depressed, if they learn to control the depression – to think positively, to do better – that would be targeting the right brain area and the right processes in the brain. I think more specificity might be achieved by psychotherapy.' (Freud would agree, having written: 'It is only the therapeutic technique that is purely psychological; the theory by no means neglects to indicate the organic basis of a neurosis.')[18] Mansuy suggests that meditation could also be a way to modulate the epigenome: 'It's the same type of idea: that someone can use their own brain to fix potential problems.' So while her own research on neuroepigenetics might eventually show the way to new, targeted chemical treatments, there are still alternative pathways to the same goal. And a great tradition of therapeutic philosophy – all the way back to our friends the ancient Stoics – could itself be a black box, which epigenetics might one day allow us to peer inside.

Even now, however, epigenetics still meets with resistance from high-profile scientists wedded to the older idea of genes as an immutable legacy, fixed at birth. Mansuy recalls the hype surrounding the Human Genome Project in the 1990s and 2000s: 'Back then people said all the time that when we understand the genome we will know everything.' But the results so far have been quite disappointing. 'In psychiatry, for instance,' Mansuy points out, 'there has been almost no progress for diseases like depression, schizophrenia, antisocial behaviours, or even suicidality as a result of genome-wide association studies. A lot of effort and money have been spent in trying to find an association between specific genes and these diseases, but so far it has yielded nothing that has helped us diagnose and treat them better.' (That said, negative results in science can be very important, as we'll see later. At least scientists now know that there is no single 'gene for' schizophrenia or many other disorders; instead there are clusters of many gene variants that seem to be associated with them. And that is a useful thing to be aware of.)

It takes time for attitudes to change. 'You still hear very dominant geneticists, in the US in particular, saying that most diseases are genetic and if we have not yet found the associated gene we will soon,' Mansuy says. It's hard, perhaps, to accept the challenge of rethinking if your reputation depends on keeping a black box closed. 'I can think of few things that would more devastate my world view,' Richard Dawkins wrote in 1982, 'than a demonstrated need to return to the theory of evolution that is traditionally attributed to Lamarck.'[19] On Twitter in 2014, Dawkins was still struggling, referring to 'the over-hyped so-called "epigenetic inheritance"'. Yet epigenetics is an expansion of genetic theory rather than a rival to it. And at the same time, there is doubtless more to be discovered in the DNA code itself, most of which is still a black box to this day. We should surely keep rummaging around in as many boxes as we can.

Now wash your hands

Some ideas have to be revived more than once in order to stick, even when the way the black box works has become well understood. Take the sad case of the young Austrian doctor Ignaz Semmelweis, who was appointed deputy head of obstetrics in Vienna's General Hospital from 1846.

The hospital had two maternity clinics, which had markedly different death rates. In the first, about 10 per cent of mothers died from puerperal fever, two and a half times as many as in the second. In fact, women entering the hospital to give birth begged to be assigned to the second clinic because the first had such a horrific reputation; some even deliberately gave birth in the street and then pretended they had done so on the way to the hospital. It was Ignaz Semmelweis alone who decided to find out why so many women in the first clinic were dying. He compared every difference between the clinics he could think of. He considered the equipment, the furnishings, overcrowding, the climate, even the religious practices of the staff. Nothing seemed like a promising culprit.

Until, that is, another doctor at the hospital, a friend of Semmelweis's, died in 1847. He had been accidentally cut by a student's scalpel while they were performing a post-mortem examination. The autopsy of this unfortunate doctor showed pathologies resembling those in women who had died of fever in the first clinic. The first clinic was also a teaching clinic for medical students, whereas the second taught only midwives. Semmelweis made the great logical leap: medical students were arriving in the first clinic to treat expectant mothers right after they had been cutting open bodies downstairs. They must, he reasoned, be carrying invisible 'cadaverous particles' on their hands that then infected the mothers. At once, he ordered that after working with corpses, the students should wash their hands in chlorinated lime solution, rather than the normal soap and water. Chlorinated lime was known as a bleaching agent that cleaned thoroughly. Semmelweis found it the most effective way to get rid of the putrid smell of dead tissue; so, he reasoned, it might be destroying the cadaverous particles along with the smell. (As we would now say, it has a disinfectant effect.) The maternal death rate in the first clinic immediately dropped by 90 per cent: in some months afterwards, they were zero.

Semmelweis began to spread this message as widely as he could. And what did he get for his efforts? Hostility and ridicule. Europe's greatest obstetricians called his view unscientific, lacking in evidence, and unsupported by any credible theory. No one knew *why* hand-washing prevented infection, so it was all too easy to decide that it didn't. What was worse, Semmelweis's idea implied that it was doctors themselves who were (unwittingly) killing their patients.

Semmelweis became increasingly frustrated and angry, writing letters that accused his opponents of being murderers. Even his wife thought he was going mad. In 1865, at the age of forty-seven, he was forcibly committed to an asylum. He died there two weeks later, probably from an infection that took hold after he was beaten by the guards. As with Lamarck, his vindication would come only posthumously, when Louis Pasteur proposed the germ theory of disease. Semmelweis had been basically right all along: his 'cadaverous particles' were bacteria breeding in the dead bodies used for student autopsies.

Astonishingly, this message has had to be revived again even within the last decade. One of the pioneers of the new movement of 'Evidence-Based Medicine' is the paediatrician Don Berwick, who co-founded the Institute for Healthcare Improvement in Boston. Berwick noticed that thousands of intensive-care patients in American hospitals were dying every year from infections after having catheters inserted in their chests. In 2004, he found an obscure study suggesting that improved hygiene by hospital staff – more frequent and systematic handwashing, combined with other practices such as using antiseptic wipes on patients – could cut the risk of such infections by more than 90 per cent. Berwick announced loudly that twenty-five thousand lives a year could be saved if these reforms were adopted immediately, and still he met with resistance and apathy: the protocols were adopted slowly and piecemeal rather than – as logic, economics, and simple humanity would seem to demand – overnight. But where they were adopted, the results were dramatic. In those American hospitals that signed up to his challenge and adopted his reforms, it was estimated that more than 100,000 patient deaths were prevented within eighteen months.[20] But too many doctors, in the 2000s as in the 1840s, continued to act like a lofty priesthood, offended by the suggestion that they were not already working optimally.

The obstetricians who ridiculed Semmelweis's handwashing complained that it wasn't supported by any reliable theory. They were correct: it wasn't. It was a black box. What no one knew was that the right theory – transmission of disease by bacteria – was just around the corner. Black boxes turn out to be very common in the history of discovery. Things may be found to work well as medicines long before the reason they work is explained by molecular biology (that ancient Chinese remedy for malaria). And the first modern steam engines were cobbled together by mechanics and inventors: a thorough scientific explanation of why they worked had to wait a century until the formulation of the laws of thermodynamics. Until then, they too were black boxes. Given what we have seen, it would be very surprising if there were not other old ideas that are right even though we don't understand how they could be right. Just because we don't know how something works, doesn't mean it *doesn't* work. So an old, discarded idea might inspire a thinker to search for the crucial missing piece that shows it works after all – just as Lamarck's writings inspired Isabelle Mansuy to proceed with the experiments that vindicated him, and just as Vladimir Kramnik went back to reanalyse the obscure Berlin Defence and then used his upgraded old weapon to beat the greatest chess player on the planet.

4

Game Changers

When innovation results from reviving an old idea in a different context – using old pieces in a new game.

'Many ideas grow better when transplanted into another mind than in the one where they sprang up.' – Oliver Wendell Holmes

Inventions are often old technologies, repurposed. The microscope came about from reversing the action of the telescope. Gutenberg's printing press employed the fundamental principles of the wine press. And finding new uses for old things is an explicit strategy of innovation in the pharmaceutical industry, where it is known as 'repositioning'. Often, it happens by happy accident. The blockbuster blue pill Viagra was originally developed as a treatment for angina. Only when the male test subjects reported an unusual side effect and firmly declined to return their spare pills did Pfizer's researcher Chris Wayman investigate the alternative use for which it was eventually marketed.[1] Another very successful medication, Ritalin, was originally engineered as an antidepressant before its usefulness as a treatment for ADHD was discovered, and it is now increasingly repurposed in 'off-label' contexts by students and other people wanting a reliable cognitive boost.

If you want to improve your own idea, there might just be an existing idea lying around that can help. Henry Ford got the notion for his car-manufacturing assembly lines from the cow carcasses hung on moving lines at Chicago meat packers. And a failed product can become a blockbuster when the game is changed. Play-Doh, for example, was originally invented in 1933 as a wallpaper cleaner. Two decades later, it was nearly obsolete and sales had fallen off a cliff. Then one of the inventor's nephews, Joe McVicker, heard from his sister-in-law Kay Zufall that the children at the nursery school she ran had just had a whale of a time playing with the company's product. It was much better than the regular modelling clay they used. Kay told Joe he should start marketing the wallpaper-cleaner goop as a toy instead, and suggested the name Play-Doh.[2] Four years later, sales of the stuff were grossing three million dollars.

Impressive things can happen when an existing product is put to a different use. And the same is true of ideas. An old idea in a new context can become very powerful indeed.

Disinformation wants to be free

One of the most ancient extant treatises on war is largely filled with advice on how to situate your armies with regard to certain kinds of landscape. 'When you come to a hill or a bank, occupy the sunny side, with the slope on your right rear,' it advises. Any place that contains 'precipitous cliffs' or 'tangled thickets', meanwhile, should be avoided entirely.[3] Most of this stuff is not especially significant for the conduct of modern warfare, even to Special Forces soldiers riding horses. And yet it became the subject of intensely renewed interest in the 1950s, from people who didn't need advice on fighting pitched battles at all.

That is because the book's most enduring message is that the greatest skill in warfare lies in victory by stealth. 'To fight and conquer in all your battles is not supreme excellence,' it insists; 'supreme excellence consists in breaking the enemy's resistance without fighting.'[4] And so the book has been consulted anew over thousands of years not by commanders of armies, but by just about everyone else. Most of its advice about the conduct of battles has been superseded, yet Sun Tzu's *The Art of War* keeps going strong.

Sun Tzu was born in 544 BCE, and became a general for King Helu of Wu, one of the pre-modern Chinese states, which lay at the mouth of the Yangtze river. He won several famous victories against the armies of other kingdoms, and his summary of his successful tactics is one of the most influential secular texts ever written. Nearly every treatise of military theory from the seventeenth to the nineteenth century is also called *The Art of War* (probably because Machiavelli's 1519 book of the same title influenced the name by which Sun Tzu's book came to be known in English).[5] But Sun Tzu's real influence lies beyond warfare per se. His book is a distillation of tips by a battlefield general, but it is his thoughts on strategy more generally that have endured, and that have been applied to almost every possible human activity in the subsequent centuries.

There was a notable cultural revival of Sun Tzu in the West in the 1980s, perhaps because the politics of the day celebrated and encouraged individual cunning. In *Wall Street*, Gordon Gekko cites him approvingly. 'Read Sun-Tzu, *The Art of War*. Every battle is won before it is ever fought.' Much later, in *The Sopranos*, Tony Soprano's therapist Dr Melfi recommends sarcastically that he should read Sun Tzu. In a subsequent session, Tony enthuses: 'I mean here's the guy, a Chinese general, wrote this thing two thousand four hundred years ago, and most of it still applies today!'[6] Bankers and mafiosi – as well as, in the interim, management theorists and business people and sports coaches – have all been enthusiastic students of Sun Tzu.

In the shadowy world of clandestine global power struggles, however, *The Art of War* is primarily considered a manual of spycraft. 'If you know the enemy and know yourself, you need not fear the result of a hundred battles,' Sun Tzu wrote. 'If you know yourself but not the enemy, for every victory gained you will also suffer a defeat. If you know neither the enemy nor yourself, you will succumb in every battle.'[7] In other words, all warfare is information warfare. This lesson was rediscovered most spectacularly by the Soviet intelligence services in the middle of the twentieth century.

In 1957, the high command of the USSR made a decision to embark on a grand campaign of strategic disinformation.[8] In everyday language, 'disinformation' usually means official propaganda directed by a government towards its own citizens; but in espionage it means something different. Disinformation is giving false information to the enemy, in the hope that they will take it to be true. (The word was in fact coined by the German High Command during the First World War.) This can be accomplished by official pronouncements and other communications that it is known the enemy is monitoring, or by more hall-of-mirrors methods. Sun Tzu recommends, for example, that a general attempt to convert the other side's spies: 'The enemy's spies who have come to spy on us must be sought out, tempted with bribes, led away and comfortably housed. Thus they will become converted spies and available for our service.'[9] One may then send them back as 'inward spies' or double agents – in the argot, you create a mole, like Kim Philby at MI6 or Aldrich Ames at the CIA, who were both clandestine Soviet operatives.

The Soviets in the 1950s wanted the Americans to think that they were much weaker than they were, and that the entire Warsaw Pact bloc was riven with disagreements, in order to make the NATO powers complacent and to conceal the progress the USSR hoped to make in establishing full communism across Europe. To create this impression required a concerted strategic disinformation campaign. The problem was that they had only one modern text on the topic: an eighty-page secret manual on disinformation by a military intelligence officer named Popov. The Central Committee ordered further research. It happened that Sun Tzu's *The Art of War* – the principles of which Mao Zedong had held dear during his long campaign in China – had been translated into Russian in 1950, soon after Mao's victory. It was then retranslated into German for publication by the East German authorities in 1957, with a foreword by the famous Soviet historian, General E. A. Razin. Sun Tzu was now a set text at East German military academies.[10]

So there appeared at this time a sudden and intense interest in the writings of the ancient master, on the part of both the Soviets and the Chinese. The grand disinformation plan certainly fits the mood of Sun Tzu's aphorisms. 'Pretend to be weak,' he advised, so that your enemy grows arrogant.[11] 'Hiding order beneath the cloak of disorder is simply a question of subdivision,' he elaborated. 'Masking strength with weakness is to be effected by tactical dispositions.'[12] And what he recommended in pitched battle, the Soviet Union sought to achieve in global geopolitics.

This, at least, was the story told by Anatoliy Golitsyn, a KGB major who defected spectacularly to the West while working at the Soviet Embassy in Helsinki in 1961. But defectors might be disinformation agents sent with a cargo of useless 'secrets' to deceive you. And that is indeed what some thought about Golitsyn. He said that the Sino-Soviet split – the worsening of relations between the USSR and China after 1960 – was simply a disinformation ruse, an apparent division to conceal an underlying unity in order to lull the West into a false sense of security.[13] But was that claim itself a disinformation ruse? Golitsyn also later insisted that the British prime minister Harold Wilson was a KGB informer – as did factions within MI5. (It turned out that the KGB had opened an 'agent development file' on Wilson, code-named OLDING, in the hope of recruiting him, but nothing came of it.)[14] On the other hand, Golitsyn also supplied information that appeared to help identify Kim Philby as a spy – which would argue in favour of Golitsyn's bona fides, unless the KGB deliberately burned Philby in order to make Golitsyn seem more credible. So was Golitsyn a plant or not?

He certainly was, according to another KGB man, Yuri Nosenko. Defecting from Switzerland in 1964, Nosenko insisted that Golitsyn was still working for the KGB and telling the Americans exactly what the Politburo wanted them to hear. Nosenko also claimed that he had personally interviewed Lee Harvey Oswald during the man's stay in Russia and decided that he was too unreliable to be recruited: so Oswald was never a Soviet agent. But the Americans didn't trust Nosenko. Instead of being welcomed to the US, he was tortured there: after being lured into a sense of security by a CIA payment of $60,000 for his information, he was imprisoned in solitary confinement in a small, windowless concrete cell, for three and a half years.[15]

This was done on the orders of James Jesus Angleton, the CIA's counter-intelligence chief. Angleton had been friends with Kim Philby: he wasn't going to get fooled again if he could help it. And Angleton had already thrown his lot in with the previous defector, Golitsyn. Hadn't Golitsyn always said that the KGB would send fake defectors afterwards to try and discredit him? So, Angleton decided, the new guy Nosenko must be a disinformation agent. To the American spies, Golitsyn's story was simply more emotionally satisfying. As Angleton's CIA colleague Pete Bagley remarked, Nosenko 'made everything sound less sinister than Golitsyn', which was evidently disappointing to those trained to find the most sinister interpretations possible. 'To me,' Bagley said, 'Golitsyn's version was simply superior.'[16]

Long after his retirement, Angleton gave a series of lengthy interviews to the journalist Edward Epstein. Why, Epstein asked him, did the decades-long 'invisible war' between the CIA and KGB really matter, when nuclear weapons threatened mutually assured destruction? Angleton shrugged and suggested that Epstein read a particular book on strategy: *The Art of War*. As Epstein later explained, the act of 'reaching into and controlling the enemy's intelligence apparatus' is what Sun Tzu called 'the divine manipulation of threads'. For the Chinese sage and modern intelligence agencies alike, this involved 'the close coordination of disinformation, which was sent to the enemy through "doomed spies" who were misleadingly briefed and then sent into enemy territory to be captured, and with the penetration of enemy intelligence by "inward spies", or moles. The eventual result was that the enemy's false picture of reality was reinforced by its own intelligence service.'[17] This, Angleton was sure, was exactly what the KGB had learned to do by studying Sun Tzu.

Angleton was fascinated by deception in nature, especially as practised by orchids: one species lures mosquitoes with a scent that smells like nectar; another has on its flower a hairy model of a female fly, to lure male flies and dust them with pollen.[18] He was also inclined to interpret events as arising from the most sophisticated kind of deception possible – which, of course, you might think was a desirable quality in your head of counter-intelligence. For this reason Angleton believed

Anatoliy Golitsyn's claim that the Eastern bloc countries were only pretending not to be ideologically monolithic. He told Epstein: 'Sun Tzu explained this strategy as succinctly two thousand years ago as anyone can today: "Hide order behind a cloak of disorder."'[19]

Others in the intelligence community, however, were not so enthusiastic about Golitsyn. In fact, J. Edgar Hoover was so unimpressed with Angleton's trust in him that collaboration and information-sharing between the FBI and CIA was brought almost to a complete halt, and Angleton was eventually forced out to spend his time breeding his beloved orchids.

And what happened to Yuri Nosenko, who was imprisoned for his warnings? It was in fact eventually accepted that he had been an authentic defector all along. In 1964, for example, Nosenko tipped off the US about an American army guard, Robert Lee Johnson, who was discovered to have been spying for the KGB in Paris; he also accurately informed his handlers that the US Embassy in Moscow was bugged.[20] His information about Oswald, too, was eventually accepted as reliable. Nosenko was given a new name and a home in an undisclosed location in the American South, and worked as a CIA consultant. In July 2008, Nosenko was visited by CIA officials bearing a ceremonial flag and a letter from Agency director Michael Hayden thanking him for his service. He died the following month.[21]

Nosenko was legitimate – but he had insisted that Anatoliy Golitsyn was a disinformation agent. So was Golitsyn a real defector or not? The archive of KGB papers brought to the West by Vasili Mitrokhin in 1991 mentions both men in official 'damage reports' after their defections, denouncing their careerism; and the KGB plotted to assassinate them both in Canada.[22] In that case, perhaps they were authentic defectors after all, each extremely suspicious of the other. Of course, if Golitsyn had been a plant, the KGB might have kept this a secret even internally, apart from a close group of senior officials who needed to know, for fear that the ruse might leak because of a mole in their own organisation. And if the KGB had managed to assassinate its own disinformation agent later, that would have proven his bona fides to the Americans beyond all doubt. Such, at any rate, are the hall-of-mirrors speculations common to this world. Whatever the final truth of the story, Sun Tzu – who taught both the CIA and the KGB how to conduct their deception operations thousands of years later – would surely shake his head in admiration.

Sun Tzu, of course, knew nothing of the hidden microphones, micro-film cameras, and hi-tech assassination methods that would be available in the twentieth century. But it turned out that his teachings on the use of espionage during ancient hot wars fought with swords and spears could be usefully re-contextualised in the global media/geopolitical world of the twentieth-century cold war. Sometimes, though, the resurrection of an old idea occurs in a different context altogether from the one in which it was originally conceived. That is why a modern business professor has gone back to one of the founders of the modern scientific method for inspiration.

Bring home the Bacon

It's a winter evening in a glass-fronted central London building. We are in a softly carpeted corporate seminar room. Suited executives are scribbling notes with the handily supplied biros branded with the logo of the University of Cambridge's Judge Business School. It's not the kind of scene where you'd expect to hear four-hundred-year-old philosophy. And yet that is precisely what is happening. The unpredictable challenges of twenty-first-century commercial life, the speaker is saying, may best be met by adopting the approach of Francis Bacon, the seventeenth-century scientist.

Professor Jochen Runde, a tall, blond bear of a man in a navy suit, takes the audience through some colourful examples of uncertainty in different commercial spheres and explains the cognitive biases that people can be prey to, and which can harm their decision-making. We tend to think that events which come readily to mind are more common than they are. This bias is called the availability heuristic. So, if there has been a horrific terrorist attack in the news recently, people are likely to overestimate the degree to which they are personally at risk from terrorism, compared to the risk of, say, driving to work. We also tend to notice only the evidence that confirms what we already think. That is confirmation bias. If someone has the idea that she knows when particular friends are about to phone her, she may remember vividly the times she is right, but forget all those times she is wrong. So her idea ends up being reinforced by confirmation bias. These kinds of bias can not only strengthen erroneous notions, but affect the way we make decisions.

How, then, should we think properly when planning for the future? Business schools talk about probability, Runde observes, but to do so is to assume you have a complete list of possibilities, so that you can assign each one a fractional likelihood, making sure they all add up to 1. In the real world, though, no one ever has such a complete list. There are always possibilities that simply haven't occurred to us. Or, to put it another way, there are always 'unknown unknowns' – a phrase Runde borrows from former US Secretary of Defense Donald Rumsfeld. In a speech widely mocked as nonsense, even though it is perfectly reasonable, Rumsfeld famously said: 'There are things we know that we know. [And] there are known unknowns. That is to say there are things that we now know we don't know. But there are also unknown unknowns. There are things we do not know we don't know.' This is true in politics, in business, and indeed in life.

Thinking in terms of probabilities, then, is all well and good. But we could always be blindsided by an unknown unknown. So, given that we cannot know everything, the million-dollar question is this: how can we intelligently estimate what we do and do not know, and then *act*? It's easy to act without understanding the state of one's own knowledge, and it's relatively easy to get bogged down in estimating probabilities without ever acting. But we want to have it all. So Jochen Runde's question is this: how is one to form a plan of action that is minimally vulnerable to being overturned by unknown unknowns?

Trying to predict, or at least be prepared for, future events falls generally into what is known as the problem of induction. Whereas *deduction* is reasoning logically from an existing set of facts (like Sherlock Holmes), *induction* is hypothesising about future facts based on current knowledge. This was already known to be an old problem in 1620, when Francis Bacon wrote:

> [A]nother form of induction must be devised than has hitherto been employed; and it must be used for proving and discovering not first principles (as they are called) only, but also the lesser axioms, and the middle, and indeed all. For the induction which proceeds by simple enumeration is childish; its conclusions are precarious, and exposed to peril from a contradictory instance; and it generally decides on too small a number of facts, and on those only which are at hand. But the induction which is to be available for the discovery and demonstration of sciences and arts must analyse nature by proper rejections and exclusions; and then, after a sufficient number of negatives, come to a conclusion on the affirmative instances . . .[23]

This has often been taken as an early statement of the modern scientific method, insofar as it first attempts to falsify hypotheses ('by proper rejections and exclusions') before claiming any positive truth. It's also the guiding principle behind what Runde calls 'A Baconian Approach to Management Decision-Making'.

This Baconian management method is disarmingly simple, and quite fun for those with a perverse imagination. The idea is first to think of the three main scenarios you can envisage as outcomes of your plan. Then you place them all on a 'favourability scale' – basically a straight line with the most favourable at the far right and the least favourable at the far left. So let's say you're going to open a new factory for your organic smoothie business. You figure the most favourable outcome is that production goes up, and increased visibility on supermarket shelves leads to more sales. Less favourable is that production goes up but there is no immediate increase in demand. The least desirable outcome is if there is a problem with production at the factory and you have to soak up the losses while it's figured out.

Now, what most people notice when they do this, Runde explains, is that the first three outcomes they thought of are all rather clustered towards the favourable end of the scale. (This is quite natural, because we are optimistic about our own plans.) So next you should try to think of a really disastrous new scenario that you can place well towards the left end of the line. Did it occur to you that the factory might be burned down by an arsonist, and also that many of your customers could be poisoned by a glitch in the production line before it was burned to a crisp?

Enjoyable, isn't it? But the key step comes next: now what you do, insists Runde, is to actively go out and try to find evidence that this most-horrible-imaginable scenario is an active possibility. (You might, say, be prompted to wonder whether anyone hates your company enough to burn down the factory, and trawl social media for nasty comments about your company by anti-smoothie folk.) When people are encouraged to do this, Runde says, they will find out some hard information that was previously not on their radar at all – in other words, an *unknown unknown* will have been found. They have to dream up the initial scenario, of course, but once they start to try to prove it, new facts arise that can now play into the decision-making. If the news is bad (there are tens of thousands of anti-smoothie activists) and this extremely negative scenario now becomes your base point, don't be discouraged. 'Then you look at the extreme positive,' Runde says – and by the same method of actively trying to confirm a hypothesis you will find out more new things. (Maybe a lot of other people are going into supermarkets and asking for your smoothie products.)

Thinking about the future can sound like woolly guesswork, but this idea is quite hard-headed. It doesn't depend on any crystal balls. 'You're not making predictions,' Runde says, 'but thinking about the possibility space.' And just by following this simple method, you could uncover unknown unknowns. 'They do this kind of thing in counterterrorism,' Runde notes. But not yet in business – or in ordinary life.

Probes

'It's quite an old idea to go back to Bacon,' Jochen Runde chuckles a few months later, seated in front of a big plate of fish and chips. 'But what we do with him is fairly novel.' We are having lunch in a Cambridgeshire pub and talking about how he and his co-author, Alberto Feduzi, decided to re-tool a Renaissance philosopher for modern business. Bacon had never been entirely forgotten, of course, but Runde and Feduzi have repurposed him. It's like dishwasher cooking. No, I had never heard of dishwasher cooking either.

Runde is very interested in repurposing in general. He has written a paper on how no one predicted that the turntable would become a musical instrument in its own right when used for scratching. (That was a cultural unknown unknown.) And cooking with a dishwasher? 'It's very good for low-temperature cooking,' Runde points out, 'because you can control the cycle, you can control the temperature, and it's also environmentally sound.' (The internet offers dishwasher recipes for such delicacies as wine-poached shrimp and strawberry cinnamon compote.) As another example of repurposing, he recalls meeting a TV producer who told him that a game-show crew had hit upon the bright idea of mounting cameras on the heads of chickens, because chickens naturally keep their heads perfectly level. Hey presto, a poultry-based Steadicam.

The repurposing of Bacon also turns out to be surprisingly practical. 'What's been interesting to me is, my work's quite theoretical,' Runde explains, 'and I've been really happy to be in a business school – delighted, because it's so interdisciplinary – but I never imagined that I'd ever, at any point, be doing stuff which might be relevant to business people!'

That's a good idea in practice, the old joke goes, but it would never work in theory. But what if a theory were, after all, eminently practical? 'I've been quite surprised, when I go and talk about this stuff to business people, to get quite positive responses,' Runde says. When he teaches executives on courses, he tells them to actually follow the procedure in class. 'Think up a novel idea, now think of something which would be devastating to your plan. They do that, and then of course the punchline isn't that; the punchline is to say to them: "Well, now do some research on that strange thing that you think might happen." *That's* when they start learning. These hypothetical things are like probes, into unknown space.' The probe may fail to find anything, and so get abandoned, 'but invariably the stuff they learn starts leading them to modify what they initially had in their business plans'.

Often we think of the imagination as working only when it wants to, in some mysterious way: as if by magic rather than focused deliberation. But this neo-Baconian method – of systematically forming hypotheses and then trying to confirm them – is actually a rigorous way of exercising the imagination in order to come up with those hypothetical probes in the first place.

The key change that Runde has made to Bacon's method takes into account our all-too-human flaws. In Bacon's recommendation for induction, Runde explains, 'you're trying to disconfirm the alternatives' – to rule out other hypotheses until only one is standing. But 'for us it's about *confirming* the alternatives. That's what we've changed, that's what we've turned around.' You don't have an idea and then try to show it's false; you have an idea and then try your best to show how it *could* happen.

The reason for this turnaround is that *disproving* alternative hypotheses is likely to be too easy because of confirmation bias. If you think of an alarming scenario (the factory poisons your customers and then is burned to the ground), the natural response is to think 'Well, of course that wouldn't happen, because our safety processes are so reliable . . .' So it would be counterproductive, in the business context, to ask people to find reasons why their hypothetical disasters would *not* happen. If you come up with a scenario just as a cheap guess, it's easy not to believe it. But if you really *try* to believe in it, you might uncover something you didn't know you didn't know. 'People will come back to you if you're the boss,' Runde says, 'and say: "Yeah we've done the job, these alternatives don't fly." It's far more productive to say to them: "I'll reward you if you give me reasons to *believe* these alternatives", because then they will really do a decent job of having a look at the information.'

So let's recap. Francis Bacon formulated an approach for solving scientific problems. For Jochen Runde, the point is not so much to solve the problem (we can't *know* what will happen in the future) as to find ways of looking at all the angles of the problem in order to know it better – including uncovering its unknown unknowns. This is a way of rethinking the process of induction. But it also has a practical application in managerial decision-making, when you reward people for trying to convince you to believe the worst. The proper incentives, it seems, can really liberate the catastrophic imagination in business.

In April 1626, snow lay thick on the ground in Highgate, north of London. According to the account in John Aubrey's *Brief Lives* half a century later, Francis Bacon was taking a coach ride with the King's doctor and suddenly had an idea: what if you could preserve meat with ice? ('It came into my lord's thoughts, why flesh might not be preserved in snow, as in salt.')[24] Bacon had to know. They stopped the coach and called in at a local woman's house, where Bacon persuaded her to kill a chicken. He then went outside and stuffed the carcass with snow. Sadly, he caught such a chill doing this that he died of pneumonia a couple of days later. Poor Bacon. He didn't see that coming.

As it turned out, the bird in question was successfully preserved, and Bacon had invented frozen food. It seems, however, that no one else at the time was interested in the phenomenon, and Bacon's one-off result was pursued no further. Eventually the pioneer of modern frozen food, Clarence Birdseye, rediscovered the idea in the early twentieth century, after witnessing an old practice that probably predated Bacon's experiment. Between 1912 and 1915, Birdseye worked as an entomologist and trapper in the frozen landscapes of Labrador, Canada.[25] He saw Inuit people hang fresh fish outside, where they were rapidly frozen solid by the sub-zero air. The fish could then be eaten, he learned, months later. Birdseye returned home and experimented with mechanical freezing methods. After a few false starts, he developed a new kind of freezer that could chill foods very rapidly, and in 1924 he established the General Seafoods Corporation, which later became the Birds Eye Frozen Food Company. The rest is history.

Birdseye was modest about his accomplishment. 'I did not discover quick-freezing,' he pointed out. 'The Eskimos had used it for centuries, and scientists in Europe had made experiments along the same lines I had.'[26] And so, after four hundred years, Francis Bacon's last, brilliant idea finally bore fruit.

The trip

How do you comfort the dying? You could try giving them LSD. It's not a facetious idea. In March 2014, Dr Peter Gasser, a Swiss psychiatrist, published the results of using LSD alongside talk therapy for twelve patients with terminal illness. It was the first controlled trial of the drug in over forty years. And it helped. The patients who embarked on carefully supervised LSD trips, Gasser reported, talked afterwards about emotions they had never expressed before, and reported feeling better – 'their anxiety went down and stayed down'.[27] One of Gasser's subjects, an Austrian social worker named Peter with a degenerative spinal condition, said: 'I had what you would call a mystical experience, I guess, lasting for some time, and the major part was pure distress at all these memories I had successfully forgotten for decades.' But he was talking about them to someone else, for the first time. Since the trial, he said: 'I will say I have been more emotional since the study ended, and I don't mean always cheerful. But I think it's better to feel things strongly – better to be alive than to merely function.' In fact, it was an old idea to comfort the dying with psychedelic drugs, but half a century of prohibition froze research to a standstill. Until now.

The British novelist Aldous Huxley, author of *The Doors of Perception*, was an early champion of acid therapy. In 1958, he wrote to Humphrey

Osmond, the physician who had introduced him to LSD and mescaline, suggesting a research project that would involve 'the administration of LSD to terminal cancer cases, in the hope that it would make dying [a] more spiritual, less strictly physiological process'. The last rites, Huxley thought, 'should make one more conscious rather than less conscious'.[28] On this subject Huxley walked the walk: on his deathbed in Los Angeles five years later, he asked his wife Laura to give him a dose of acid. She did so, staying at his bedside and talking to him for hours until he breathed his last. In a remarkable letter to his brother a few days later, Laura reported that it had been 'the most serene, the most beautiful death'. And then she ended her loving report by asking a question. 'Now, is his way of dying to remain our, and only our relief and consolation, or should others also benefit from it? What do you feel?'[29]

Such ideas in the popular culture of the 1950s and 1960s were the humane, optimistic version of the general fascination with hallucinogens. The dark side of that fascination is exemplified by the CIA's notorious MK-ULTRA program, which according to a 1985 Supreme Court judgment describing it 'was established to counter perceived Soviet and Chinese advances in brainwashing and interrogation techniques', and concerned itself with 'the research and development of chemical, biological, and radiological materials capable of employment in clandestine operations to control human behavior'. In other words, they performed illegal human experiments, which included giving unwitting subjects doses of LSD.[30]

But in the late 1960s, LSD and other hallucinogenic drugs were outlawed in the US and the UK, effectively rendering clinical experiments impossible – lawful ones, at least. According to many psychiatric pharmacologists, this set back research into psychiatric therapies for decades. Only after half a century is scientific research into the effects of LSD and other psychedelics such as psilocybin (the active ingredient of magic mushrooms) picking up again.

Why? Well, for one thing, depression is now the leading cause of disability on the planet, according to the World Health Organization.

And there is a need for better medications than those so far provided by the pharma industry. As Dr Robin Carhart-Harris, a researcher in neuropsychopharmacology, told the writer Ed Cumming: 'Given the magnitude of the problem, there's huge potential that psilocybin, and maybe other psychedelics, will be a big help.'[31] Depression is not the only condition potentially treatable by these once-forbidden compounds. They seem to be of potential help in anxiety and nicotine or alcohol dependence as well. According to one intriguing scientific meta-analysis published in 2012, 'A single dose of LSD, in the context of various alcoholism treatment programs, is associated with a decrease in alcohol misuse.'[32] A similar approach might work for those who love cigarettes too much: 'Psilocybin may be a potentially efficacious adjunct to current smoking cessation treatment models'.[33] Even the ordinary recreational use of psychedelic drugs in the American population, it seems, is associated with reduced psychological stress, and less risk of suicide.[34] The studies are mounting up. None is definitive, but the implications all point the same way. The potential benefits of these substances, many physicians and researchers now say, demand a change in their legal status.

So why do psychedelics seemingly work in therapeutic interventions? There is an intriguing hypothesis, which seems quite compatible with the teachings of the tradition in psychology that runs from the Stoics to CBT. Brain-imaging studies performed on patients who have taken psychedelics show a measurable, concrete phenomenon: a bio-electrical reality underlying an age-old metaphor.

We often say that people who are depressed or addicted have thoughts that run in certain well-worn grooves – psychological 'ruts' that repetition only strengthens. How do you get out of such a rut? As we have seen, the psychological tradition recommends that we try to recognise repetitive bad thoughts, to disrupt our patterns of automatic thinking. What psychedelics do is to actively disrupt the overall organisation of neuronal firing patterns in the brain. (This is known as 'broadband cortical desynchronisation'.)[35] That sounds scary, but for some people it's just what is needed. If a 'rut' is an entrenched pattern of neuronal firing, then something that disrupts patterns will help smooth it out or reset it. 'What you see with psychedelics is this dismantling of organisation, a scrambling in the cortex,' Carhart-Harris explains. 'These drugs introduce a kind of storm, but in the context of treating a pathology, it can be a useful storm, a reboot of the system.'[36]

A similar effect, as it happens, may be the benefit of another treatment long thought consigned to the annals of medical savagery – ECT (electroconvulsive therapy), aka shock therapy, as endured by Jack Nicholson in *One Flew Over the Cuckoo's Nest*. ECT never actually went away as an option for the most severe cases of depression, though voltages have been decreased, and the use of anaesthetics and muscle relaxants has been brought in. A modern reinvention of ECT being trialled by the American psychiatrist Sarah Hollingsworth Lisanby uses magnetic stimulation instead of direct electric currents. It is called Magnetic Seizure Therapy, which may not be the most inspiring brand name.[37] Still, it seems to help people with particularly severe depression, just as old-style ECT can, for all that it was employed abusively in the past. ('Depression kills, while ECT saves lives,' Lisanby insists.) And the theory is that this is because of a similar effect to that attributed to psychedelics – a kind of 'reboot' of the dysfunctional brain network. Our age is very eager to use computer metaphors to describe the brain, and to do so is often conceptually reckless, but this one at least seems to be a good analogy. The brain is an electrical network, and it appears that psychedelics or magnetically induced seizures can produce desirable changes in experience by following the first piece of advice offered by tech-support people the world over. Have you tried turning it off and then on again?

The resurgence of therapeutic drug research has been called the Psychedelic Renaissance, but LSD was not invented for the purpose of alleviating depression. It seems obvious now that consciousness-altering drugs might change the experience of people with a troubled consciousness. But the Swiss chemist who discovered the trippy molecule lysergic acid diethylamide was looking for something else altogether. Albert Hofmann stumbled upon LSD in 1938 while researching possible medications to stimulate breathing and circulation. It was no good for that purpose, so Hofmann set it aside. Only five years later did he decide to re-examine it, and while doing so he accidentally got some into his system. He liked it so much he took it again deliberately. And so began the psychedelic age.

Hofmann thought LSD was 'medicine for the soul', just as many modern psychiatric researchers suspect it might be. But that potential was only discovered once it had been repurposed – taken out of its original context as a possible stimulant, and reconsidered for recreational intoxication, and then therapy. So, too, Sun Tzu's classic treatise *The Art of War* is mostly about the minutiae of pitched battles on different kinds of territory, but it has been repeatedly repurposed as a classy self-help book for bankers, gangsters, and spies. And Francis Bacon's method of induction has been repurposed to help business people identify the worst, so they don't fear it. If an idca looks moribund or simply irrelevant, perhaps it just needs to be slotted into an entirely new game to come alive once more.

5

Are We Nearly There Yet?

An old idea might be considered viable only when attitudes change.

'A new scientific truth does not triumph by convincing its opponents and making them see the light, but rather because its opponents eventually die, and a new generation grows up that is familiar with it.' – Max Planck

Some ideas are simply ahead of their time. When they are first proposed, the prevailing cultural, social, or economic order is incapable of embracing them: they pose a challenge to powerful interests, or don't seem necessary yet, or simply represent too startling a conceptual change. This is common in politics and ethics: democracy is revolutionary ferment before it is parliamentary routine, and it took a long time to abolish slavery. Yet equally stiff resistance can also greet concrete ideas in science, technology, and habits of dining. An idea might be too radical, or too threatening, or too plain weird to be accepted in a particular cultural context. Or it might just come from the wrong kind of person. The idea's originator may be able to make some headway against such prejudice in her own lifetime, or it may take hundreds or thousands of years for vindication to arrive. And too long a wait could be not just disappointing but dangerous.

Revape

The Chinese pharmacist Hon Lik used to smoke three packs of cigarettes a day while working in a laboratory that formulated health products based on ginseng. But then his father, who also smoked like a chimney, died of lung cancer. In the early 2000s, Hon himself managed to quit. And then he started thinking. Surely there must be a safer way for people to enjoy the pleasures – and the well-attested cognitive benefits – of nicotine?

Hon rethought the cigarette. What if there was a way to deliver both a nicotine hit and something approximating the satisfaction of inhaling smoke, without the tar and noxious carcinogens? That would be a much better replacement for ordinary cigarettes than gum or patches.

He hit upon the idea of using a lithium battery to power a heating element that would instantly vaporise a solution of liquid nicotine mixed with propylene glycol, the stuff that makes 'dry ice' in nightclubs and which is also approved as a food additive in most countries. The resulting 'smoke' was basically nicotine-laced water vapour. Hon had invented the e-cigarette. No tar, no known carcinogens, no known danger from second-hand inhalation. It was 2003.[1] Hon patented his invention and electronic cigarettes gradually became popular, first in China and then in the West as well.

But by 2014, the backlash was well under way. The EU was considering imposing harsh regulatory and tax restrictions on electronic cigarettes, and their use was gradually banned in more and more public places in America: in public parks as well as bars in Los Angeles and New York City. Vaping – the verb coined by enthusiasts – was being treated exactly like smoking. Officials cited fears that vaping was a 'gateway' to smoking. Actually, research suggested the opposite: smokers were using electronic cigarettes in order to stop smoking. In 2015, the UK research body Public Health England declared that, according to the 'best estimate so far', vaping was 95 per cent less dangerous than smoking.[2]

Some public-health experts decried official anti-vaping policies in startling terms. In Britain, Professor David Nutt pointed out that if more and more smokers switched to vaping, deaths from lung cancer would be vastly reduced. This meant, he said, that electronic cigarettes were 'probably the most significant advance' in public health since antibiotics.[3] Imagine how many lives could have been saved if vaping had caught on earlier. And it could have done. Because the basic idea for an e-cigarette long predates Hon Lik's invention: it is actually half a century old.

A design for what looks strikingly like a modern e-cigarette was patented by the American Herbert A. Gilbert, in 1965. In a 2013 interview, Gilbert remembered: 'The problem, as I concluded, was that when you burned leaves and wood, even if you did it in your back yard, it yielded a result that no one wanted to take into their lungs . . . Using logic I had to find a way to replace burning tobacco and paper with heated, moist, flavored air.'[4] His patent for a 'smokeless non-tobacco cigarette' emphasises its health benefits: 'Persons who wish to smoke but have been advised against such a practice by their doctor may use this invention to maintain the satisfaction of smoking without any of its disadvantages.' The patent even anticipates the range of e-juice flavourings available to today's vapers: the contents may vary, it points out, 'from slightly mentholated water to a solution which would simulate artificially the flavor of Scotch whisky. Many other solutions and flavors may be employed.'[5]

But Gilbert's invention was never put into production. 'I showed it to chemical companies, pharmaceutical companies and tobacco companies,' he said, 'and they did what they did to try to protect their markets.' He was shown the door wherever he went. That old paranoid story about how Big Tobacco suppressed the idea of a 'safe cigarette' seems, in this instance, to contain a germ of truth. 'Timing can be everything,' Gilbert points out, 'and I was ahead of my time, and in the midst of what some might say was the most powerful advertising period of big tobacco.'

In this way social and cultural forces can suppress an idea that only emerges successfully when attitudes have changed. As late as 1960, after all, just a third of American physicians believed that smoking caused cancer.[6] The 1964 report by the Surgeon General made the link clear, but it took time to percolate into public opinion. In that historical moment there was no great urgency to produce a safer alternative. By the twenty-first century, on the other hand, the health risks of smoking were uncontroversial, tobacco companies had been the subject of epic class-action lawsuits, and indoor smoking bans were being enacted all over the world. The moment was right for the reinvention of the electronic cigarette. Its time had come.

It's not just the self-interest of industry that can doom an idea to the shadows. One of the best ideas anyone has ever had was denounced on principle by one of the greatest philosophers in history. It was eventually vindicated only two and a half thousand years later.

Atoms of desire

What happens if you keep cutting something in half? You can slice a loaf of bread in two, and then repeat the operation again and again. When you have nothing left but crumbs, you can take a finer knife and bisect those. At some point the crumbs will be too small to cut any further. But what if you had godlike vision and an infinitely thin blade? Could you keep cutting for ever?

The Athenian thinker Democritus was known as 'the laughing philosopher' for his constantly amused attitude towards human folly. He also happened to be one of the great polymaths of the age, having travelled far and wide in Egypt, Babylonia, and Persia to learn all he could. After those adventures, he preferred solitary study, locking himself up in a little shed in his garden, and occasionally venturing out to visit tombs.[7] The catalogue of his works at Alexandria listed more than sixty treatises on everything from physics and medicine to botany, musical theory and painting, though almost nothing survives. (Possibly it helped this extraordinary intellectual productivity that Democritus disapproved of sex, on the grounds that sex was so much fun it overwhelmed the mind.)[8]

What we do know is that, using nothing but his powers of reasoning, Democritus decided that the idea that you could keep cutting something for ever was absurd. Eventually, he concluded, you would hit something fundamentally indivisible. Something that could not be cut in two. He called this 'that which cannot be cut', or in Greek *a-tomos*. The atom.

Everything we see, Democritus decided, had to be made of atoms. But another ingredient was needed for his picture of fundamental reality. Evidently, things around us change. A loaf of bread goes mouldy. A puddle dries up. This change, Democritus reasoned, must be the result of atoms being added to, or taken away from, those objects. Say a servant is coming into your room with a bread-based snack. Well, you can already smell the bread before it arrives at your table. This must mean that some atoms of bread get to you before the whole loaf does. Therefore reality, at bottom, cannot be completely packed with atoms, or there would be no way for them to move and exchange places with one another. So there also had to exist empty space for the atoms to move in. Democritus called this empty space the void. 'All that exists is atoms and the void,' he wrote. 'Everything else is mere speculation.'

The interactions of atoms, Democritus thought, must also fundamentally explain the workings of living organisms, and even the human mind. This was a dangerously radical idea. But he insisted: it was only by 'convention' that people described things as hot or cold, sweet or bitter, or as having certain colours. In reality everything was atoms colliding in the void. There was no place in this picture for divine intervention or design. And there was no place for the soul. As the quantum physicist Niels Bohr later pointed out, Democritus's ideas would have seemed to his contemporaries to be 'extreme materialistic conceptions' with a 'fantastic character'.[9] His culture simply was not ready for them. Not long after Democritus and his fellow Atomists (the first may have been his teacher Leucippus: it is not clear exactly which ideas can be ascribed to which thinker) proposed it, the atomic hypothesis was denounced by the philosopher of antiquity who exerted the greatest influence on the next two millennia of Western civilisation: Aristotle. He insisted that atomism was unthinkable because it seemed to undermine the divine unity of the world – for one thing, Aristotle maintained, the elements had to be smoothly continuous substances rather than jostling collections of tiny bits – and it seemed to leave no room for the idea of purpose in nature (an idea which we shall revisit). In sum, atomism was unacceptably random and severe. Plato, for his part, is said to have wanted to burn all Democritus's books, but was unable to because they were already in wide circulation. He contented himself with never referring to Democritus in any of his own texts. And so atoms were more or less suppressed, and physics was frozen in time.

Long afterwards, Bertrand Russell was to pronounce that philosophy went downhill after Democritus and did not recover until the Renaissance. (One important factor, as argued in Stephen Greenblatt's *The Swerve*, was the rediscovery at the beginning of the Renaissance of the philosophical poem *De rerum natura*, by the atomist Lucretius, who lived four hundred years after Democritus.) In the twentieth century, the physicist Richard Feynman said that if he had to choose a single scientific one-liner to pass on to a future post-apocalyptic civilisation, it would be the atomic hypothesis. So much flows from it. And Democritus was simply right: at least, according to our current best theories, there really are ultimate, indivisible particles of matter. They're not our 'atoms' but the smaller constituents of them, known as quarks and leptons. This is just a historical accident: the early-nineteenth-century chemists who discovered atoms, foremost among them the Englishman John Dalton, really thought they had found the fundamental parts of matter, and so christened them atoms in honour of the Greek. ('Needless to say,' notes Thomas Kuhn sardonically, 'Dalton's conclusions were widely attacked when first announced.')[10] What had Democritus done, so long ago? He had just thought very hard about the problem, and arrived at the correct answer. That is an example of armchair thinking – or rationalism, an idea to which we shall return – at its finest.

Some moderns, though, are less impressed than Feynman or Russell were by Democritus's feat of reasoning. The Nobel-winning physicist Steven Weinberg, for one, is rather sniffy about it in his history of science, *To Explain the World*: 'I think one should not overemphasize the modern aspects of Archaic or Classical Greek science.' Of Democritus in particular, he complains: 'Nowhere in the fragments of his books that survive do we see any effort to show that matter really is composed of atoms. Today we test our speculations about nature by using proposed theories to draw more or less precise conclusions that can be tested by observation. This did not occur to the early Greeks, or to many of their successors, for a very simple reason: they had never seen it done.'[11]

There is a sin in the academic history of science called 'presentism', which according to its opponents errs in judging the past by the standards of the present. (This is an aspect of 'Whig history', which assumes that past generations were struggling all along to become as wonderful as we are.) Here, as throughout his book, Steven Weinberg is knowingly committing presentism, in criticising Democritus for not performing experiments. But to refuse any hint of presentism, conversely, is to be unable to give unfairly maligned thinkers their due. It is only by the standards of the present that we know Democritus really was correct after all. Yes, as Weinberg points out, it is true that he did not grow up in a culture where scientific experimentation was the norm. Nor was there available the kind of finely calibrated scientific apparatus that could have allowed him to test his theory. Even so, he was right. But the culture he found himself in was not ready for his idea in two ways. First, its most influential thinkers held to principles such as souls and purpose in nature which seemed to be undermined by his fantastically minimalist world of atoms and void. Second, Democritus's culture was simply not physically equipped to allow him to demonstrate the truth of his ideas.

Atoms were not generally accepted until surprisingly recently. Even some of the most prominent scientists of the late nineteenth and early twentieth centuries refused to endorse their existence. They had what seemed to them a pretty good reason: no one had ever seen one. 'Almost to our day,' Niels Bohr wrote in 1954, atomic ideas 'were regarded as essentially hypothetical in the sense that they seemed inaccessible to direct confirmation by observation because of the coarseness of our sense organs and tools, themselves composed of innumerable atoms.'[12]

The controversy began to settle down only after the 1905 paper on Brownian motion by Albert Einstein, which interpreted the jiggling-about of particles in a liquid as the particles bouncing off individual molecules. This seemingly put paid to the anti-atomist conception of a substance as a natural continuum, which could not be separated into component parts. Yet some eminent scientists still held out. They were strict empiricists, refusing to countenance the existence of phenomena that could not be confirmed by the senses. So you could accept that a liquid was hot; but the hypothesis that the heat was the effect of invisible atoms moving around in the liquid was at best unproven. The great physicist Ernst Mach, for whom the speed of sound is named, argued that, useful though the atomic theory might be, you were not allowed to believe in the reality of things that could not be directly observed. Atoms were for him instead 'mental artifices' – or, you might say, mere ideas. Another holdout, though, caved in 1908: after further experiments that supported Einstein's interpretation, the chemist Wilhelm Ostwald declared that Democritus had been right all along: 'I am now convinced that we have recently become possessed of experimental evidence of the discrete or grained nature of matter, which the atomic hypothesis sought in vain for hundreds and thousands of years.'[13]

It's worth emphasising the point, however, that Mach and Ostwald both had respectable philosophical reasons to doubt atoms. 'We are inclined now to regard Mach and Ostwald as pedants on this point,' observes the historian of science Philip Ball, 'but it would be more appropriate to say that they were merely expressing suspicion of the occult.'[14] The controversy in science about whether 'occult' causes – things we cannot directly perceive – should be accepted as explanations was a long-standing one. Many early critics of Isaac Newton had complained, not unreasonably, that in positing gravity as something that could act instantaneously over vast distances, Newton was being unscientific: he was reviving the genre of 'occult forces' that early scientists were determined to stamp out.[15] Eventually – but only in the twentieth century – scientists generally accepted the invisibly small as real. That was itself the result of a long cultural shift. And only once it was accomplished could Democritus be given his due, at long last.

It's just not crickets

When I was a student, a friend gave me a copy of an old pamphlet she had picked up in a secondhand bookshop. Its title was *Why Not Eat Insects?* How splendid, the confidently bluff Victorian phrasing of a question that seemed to have a very obvious riposte – *why?* I remember vividly my disgust at one of the recipes in particular, which was straightforwardly named 'Moths On Toast'. And yet the author of this worrying pamphlet was way ahead of his time. At one British chain restaurant serving Mexican street food, crickets are now on the menu and selling thousands of plates a week.[16] Another insect-only eaterie is called Grub Kitchen. And in 2014 the UN organised a global conference on insect-eating as a way to save the world. Has this idea's time finally come, too?

Why Not Eat Insects? was originally published in 1885, by an optimistic Englishman named Victor M. Holt who wanted to persuade his compatriots of the virtues of an entomological menu. 'There are insects and insects,' Holt explains in the introduction. 'My insects are all vegetable feeders, clean, palatable, wholesome, and decidedly more particular in their feeding than ourselves. While I am confident that they will never condescend to eat us, I am equally confident that, on finding out how good they are, we shall some day right gladly cook and eat them.' Today some of his countrymen are beginning to do just that.

Of course, people in non-European cultures have long eaten insects. It is estimated that two billion around the world do it today. The reason it never caught on in Europe, it is thought, is that the temperate climate means that there are simply fewer and smaller insects around to catch easily, while the success of larger-animal husbandry just rendered insects irrelevant as an alternative food source. Yet elsewhere they are still traditional delicacies: in Japan one can find canned bee larvae (hachinoko), and fried grasshoppers (inago). And in other places they are unremarkable bar snacks: in Malawi, you might be offered chilli-and-lime termites to crunch on with your beer – which is what gave entrepreneur Shami Radia the idea to found Grub, a London-based company supplying edible insects to restaurants.[17]

Victor Holt pointed out the global examples too, albeit in a manner fruitily redolent of your average nineteenth-century Englishman. 'We imitate the savage nations in their use of numberless drugs, spices, and condiments,' he wrote; 'why not go a step further?' He was optimistic that his compatriots would eventually see sense. 'I foresee the day,' he wrote, 'when the slug will be as popular in England as its luscious namesake the Trepang, or sea-slug, is in China, and a dish of grasshoppers fried in butter as much relished by the English peasant as a similarly treated dish of locusts is by an Arab or Hottentot.'

But why? Holt gives several reasons why his readers should alter their diets. First, he advises sagely, 'philosophy bids us neglect no wholesome source of food'. Why waste what is provably nutritious? His second reason might have held less sway with the objects of his encouragement. 'What a pleasant change,' he enthuses, 'from the labourer's unvarying meal of bread, lard, and bacon, or bread and lard without bacon, or bread without lard or bacon, would be a good dish of fried cockchafers or grasshoppers.' It would certainly be a *change*, but labourers might have been forgiven for not thinking it a pleasant one. Holt suggests, too, that the harvesting of insects for food would reduce the number of insects that might eat farmers' other crops. And then there is simple prejudice of the palate: why should those *bon vivants* who scoff raw, living oysters turn up their noses at boiled caterpillar larvae? 'It may require a strong effort of will to reason ourselves out of the stupid prejudices that have stood in our way for ages,' Holt writes, 'but what is the good of the advanced state of the times if we cannot thus cast aside these prejudices, just as we have caused to vanish before the ever-advancing tide of knowledge the worn-out theories of spontaneous generation and barnacle geese?'

It took quite a long time, but these prejudices are at last under serious pressure in the modern West. Grub Kitchen promises in its promotional materials to introduce adventurous gastronomes to 'the exciting new practice of eating insects (entomophagy)'. The practice is actually as old as the hills, of course. Moses approved of eating some insects; John the Baptist was said to live on locusts and honey in the desert; and the Romans feasted on fattened grubs. But the modern argument in favour of entomophagy has a different emphasis from Holt's. The new insect-eaters point to the 'protein challenge' – given billions more people on Earth in the coming decades, how will everyone get enough protein? Worries about the ability of the land to support a growing population go back a long way, as we shall see later on in this book. The modern fans of entomophagy address this concern, though, with a particularly modern language – that of environmental 'sustainability' and 'food security'.

The world's first global conference on edible insects took place in May 2014, under the auspices of Wageningen University in the Netherlands, and the UN's Food and Agriculture Organization. It was called 'Insects to Feed the World'. 'The basic principle is that insects will generate a sustainable solution to the increasing demand for animal protein caused by an ever-growing world population,' the conveners announced.[18] Cows and other livestock animals are notoriously inefficient at converting vegetable matter into protein for humans: they need an awful lot of water and land. Insects obviously require much less. They are even, it is said, a 'superfood', packing a lot of nutrition into a small space. And so, in a charming attempt to domesticate an idea that still puts a lot of people off, edible insects are now referred to as 'mini-livestock'.[19] One thinks immediately of a bonsai cow or foot-long pig, but no: these mini-livestock have six legs and antennae. Others, alternatively, speak of 'harvesting' insects, as if they're not really animals at all. Clever renamings help: waxworms are not called waxworms by people selling them as food because 'worms' just sounds horrible;

instead they are 'honey bugs' or 'honeycomb caterpillars'. One entomologist thinks we should call insects 'land shrimp'.[20]

To overcome the disgust of many fortunate people who already have enough non-insect food to eat, however, worthy words about sustainability and food security will probably not be enough. A massive PR effort is probably needed, encompassing subtle shifts of vocabulary and more direct interventions such as pop-up restaurants and perhaps even, as suggested by some conference attendees, insect aisles in the supermarkets.[21] It's not so long ago, entomophagy enthusiasts point out, that people in Britain and the US felt weird about the idea of eating raw fish; but now sushi is extremely popular. They say the same will happen with insects – if the idea is properly marketed to 'foodies', people who are constantly yearning to put something new into their mouths. The legendary Nordic Food Lab, founded by cook René Redzepi, was at the UN conference serving insect-based snacks, and its researchers have developed recipes including moth mousse, cricket broth, and roasted locusts. These Nordic gastro-adventurers are not above employing the natural fallacy, either: in a communiqué about the different flavours of ants available, they point to chimpanzees' eating of termites and write: 'Eating insects like these is part of our evolutionary heritage.'[22] (Well, yes, if you like. So is swinging from trees and being unable to do the crossword.) Others have decided to try to wean squeamish Westerners on to insects by disguising the origin of the product: thus the American Company Six Foods sells Chirps, which are a bit like potato chips except that they're made from cricket flour. But Laura D'Asaro of Six Foods would like for everything to be explicit and above-board: she'd love us to be able to just walk into a fast-food joint and order an 'ento' burger.[23]

There is even, some argue, a moral obligation on those of us who don't eat insects to start doing so. Adena Why, an entomologist who spoke at the Insects to Feed the World conference, told a reporter: 'Ethically, especially as Americans, we have a responsibility not to consume as many resources as we do, and if entomophagy is the way to do it, then the faster we do it the better.'[24]

The disgust reaction of people unaccustomed to eating insects seems a real barrier. It's possible that the two-thirds of the world's population who don't currently eat insects might settle, this time, on a kind of halfway house. If experts consent, it might well make sense, at least, to use insects as animal feed in the meat industry – giving, as you might say, mini-livestock to maxi-livestock. Chickens, the FAO points out, will naturally eat worms and larvae from the ground. And insect exoskeletons contain a substance called chitin, which it is thought might strengthen food animals' immune systems and so decrease the use of antibiotics in them that is helping to produce the scourge of antibiotic resistance.[25] Now that might be a good definition of 'sustainable'.

The call for more people to eat insects, on the other hand, may return again and again without ever really sticking. Even so, this example does show how an old idea can seem more viable simply through the passage of time and the concomitant change in social attitudes. When Victor Holt published *Why Not Eat Insects?* in 1885 it could have been looked upon as an amusing eccentricity that obviously would never catch on in England. No one in late-nineteenth-century Britain thought that eating insects was necessary. In 2014, though, the speakers at the Insects to Feed the World conference did argue explicitly that eating insects might be the only viable way to feed future populations. And because of such modern concerns a global UN conference was organised to discuss an idea that to many people is no longer a whimsical piece of exoticism but an urgent ethical call to arms.

The Admiral and the iPhone

Elana is a young interactive designer, originally from Uzbekistan, who moved to London as a teenager and now works for a hip media agency in Shoreditch. Over a drink one summer evening, she tells me that she and a couple of friends are working on a side project to enable programmers to write their programs in Russian, or French, or even Chinese. As things stand, all modern programming languages are based on

English (commands like 'if', 'then', etc.). 'Why should it always have to be English?' Elana asks.

Not that she is virulently anti-anglophone. But she thinks it would be cool if coders were able to use their native language when talking to computers.

'That's a good idea!' I say.

It turns out that it's also an old idea.

Today, we celebrate the intuitive usability of tablets, phones, and laptops. Computers are designed to be accessible to all. Yet half a century ago, the hi-tech community wanted to protect the fiendishly complex way things had always been done, and aggressively resisted proposals to democratise computing. There might be no iPhone and no social media today had it not been for one rebel pioneer: the legendary Grace Hopper.

During the Second World War, the Harvard Computation Laboratory housed early computers used to solve problems in ballistics. These machines had to be programmed by literally plugging in and unplugging hundreds of physical cables. Grace Hopper was a young lieutenant in the US Navy assigned to that lab. Her group also included the soon-to-be-famous mathematician John von Neumann, and together they laid down some of the fundamental principles of computation. (Some historians actually argue that von Neumann took credit for Hopper's ideas at the time.)

But it was after the war that Hopper had her big idea. By then computers were programmed in machine code, which to the untrained eye looks like hexadecimal gibberish – B0 61 E3 79, that sort of thing – but which was still an advance on a forest of plugs. Hopper wanted to make programming even easier. What if the user could type instructions into the computer that were easier to memorise than pure machine code? And what if a program then automatically performed the tedious work of translating those instructions into a language the machine could understand? That would be a kind of *automatic programming*. Then, Hopper dreamed, programmers would be freed from the hundreds of hours of banal arithmetical manipulations required to write machine code, and they could concentrate on a more high-level, creative view of what the program was intended to do.

Hopper didn't just dream: she made it happen. In 1951, she created the first 'compiler', a program that would translate a new, more human-friendly set of instructions into machine code automatically. This was a vastly more efficient system. In one test, the old-fashioned way of writing a program to solve a simple geometrical equation took three people more than fourteen hours, of which a full eight hours was spent just on translating the program laboriously into machine-readable instructions using a code manual. The same equation was turned into a functional program using Hopper's compiler by a single person in less than one hour.

Did the programming world fall over itself to hail Grace Hopper for her brilliant revolution? No. Quite the opposite, in fact. The head of computer operations at General Electric, Herb Grosch, led a vocal resistance movement that for years argued that programming was far too delicate and ingenious an activity for any part of it to be left to the computer itself. It was clear to Hopper that programmers regarded themselves, as she put it, as 'high priests', jealously guarding their status as intermediaries between ordinary people and the occult computer brain, and felt that her compiler was a threat. Meanwhile, attempting to persuade senior management of the virtues of her idea at the company where she worked, the Eckert-Mauchly Computer Corporation, was even more difficult. Hopper recalled: 'It was obvious to everyone that computers could only do arithmetic; they couldn't write programs. And no matter how much you explained they weren't really writing a program, they were only piecing one together, people just didn't understand that.'[26]

Hopper was eventually successful in persuading her next employer, Remington Rand, to set up a new department for research into automatic programming. But the (now-legendary) programmer John Backus, one of Hopper's few allies, recalled that even by the mid-1950s the general feeling was: 'It was obviously a foolish and arrogant dream to imagine that any mechanical process could possibly perform the mysterious feats of invention required to write an efficient code.'[27]

The tide was turned eventually when Hopper decided to abandon trying to persuade the programmers themselves. Instead she appealed to business executives, the potential users of her product, with promises of greater speed, efficiency, and ease. One of her early compilers was called 'Business Language version 0', which reflected her idea that programming could be made transparent and simple enough for business people to understand. Hopper wanted to democratise computing and championed programming dialects that resembled natural language – which is why mainstream programming languages today contain normal English words such as 'if' and 'stop'. Her strategy worked. By the end of the decade, payroll and inventory applications were being written at various military departments and industrial corporations with Hopper's own language, FLOW-MATIC – a system that hardcore programmers still mocked because it was too easy to understand.

The majority of programmers were clearly too close to this innovation to see the need for it – they had so much invested in the ordinary way of how things were done that they couldn't appreciate the virtue of Hopper's idea for widening the scope and application of computers in general. They were too hung up on the established *process*; whereas Hopper had a vision of an *outcome* that few others could see. And it is only because Hopper and her handful of allies persevered throughout those ten years that coders on internet forums today can offer excruciatingly detailed smackdowns to the naïve among their number who think that Python really is a good programming language. Hopper invented the very *idea* of a programming language. And it was only because she forced this revolution in computer accessibility that, later, non-mathematicians could start playing around with ideas such as the graphical user interface. So: no Hopper, no iPhone. In recognition of her extraordinary achievements, Grace Hopper was later promoted to Rear Admiral in the US Navy, but she is hardly a household name. And yet she invented a crucial technique that helped build the world around us.

But even if a big idea eventually succeeds, other good ideas proposed alongside it can be lost for ever. Another innovation that Hopper proposed during this era seems extraordinary now only because it never took off.

In her early thinking about programming, Hopper had realised that the person who writes a compiler is in effect also a linguist. She is free to choose the structure and syntax of the new programming language, which can work any way she thinks will be helpful and flexible. The compiler itself will always translate that language back into the arcane numerals that feed the computer's brain. So it doesn't even matter what real-world language you base the compiler on in the first place. Why not give the user several options?

So, for one demo to management, Hopper wrote a compiler that allowed you to program not just in English but, if you preferred, in French or German: the system would automatically translate any of the three input languages into the same machine code. Thus, typing 'IF GREATER GO TO OPERATION 1' was functionally equivalent to typing 'SI PLUS GRAND ALLEZ À OPÉRATION 1' or 'WENN GRÖSSER GEHEN ZU BEDIENUNG 1'. Her managers were aghast. To them, she said later, it was perfectly obvious that 'an American computer built in blue-belt Pennsylvania couldn't possibly be programmed in French or German'.[28] Hopper had to promise management that from then on the program would only accept English input.

So died a small dream of cosmopolitan computing. Only to be reborn half a century later in East London. Perhaps it's another idea whose time has come.

Grace Hopper always had an impressive disrespect for mere convention. In the late 1960s, she became Director of the Navy Programming Languages Group at the Pentagon. On her arrival, she saw that no one had thought to furnish her new department. So, one night, she and her colleagues 'liberated' the furniture they needed from other offices. When someone complained, Hopper simply pointed out smilingly that the desks and chairs hadn't been nailed down.

In her own office she kept a pirate flag, and a strange clock that ran anticlockwise. It was a perfectly functional clock once you learned to read it, but it was not the way clocks usually worked. Why display such a weird instrument? To combat the fact that human beings are allergic to change. 'They love to say, "We've always done it this way,"' Hopper explained. 'I try to fight that.'[29] And so she did.

Hopper's polyglot compiler is what we might call a power-up idea. Power-ups in video games are things your character finds that make you more powerful by giving you extra abilities: a new weapon, increased health, or a more powerful jump. Some ideas are like that too. Being able to program computers in any language would be empowering for those whose first language is not English. Similarly, Jochen Runde's method of trying to find evidence for disastrous outcomes could be an empowering tool for the entrepreneur.

Of course, if you are offering to give power to one group of people, you might be threatening to take it away from another group. Often, then, power-up ideas will meet resistance. And so the 'high priests' of early programming were obstructive when Hopper introduced her first compiler. Even just in English, it was a power-up that she hoped could democratise computing. But if a business person could program a computer in something that looked like English, then a programmer who was a wizard at machine code would understandably feel threatened. It would be in programmers' interests for the idea not to catch on.

And sometimes people just don't like the messenger. If you suspected that the initial resistance to Hopper's brilliant innovation was exacerbated by the fact that she was a woman, you'd be right. In 1960, Hopper participated, along with several other women computing pioneers, in the development of the computing language COBOL (Common Business-Oriented Language). According to one historian of technology, numerous vocal sceptics of this latest new language at once 'concluded that the fruits of such an unstructured, female-dominated process could not be expected to survive, let alone flourish'.[30] By the year 2000, notwithstanding such negative prophecies, it was estimated that COBOL constituted about 240 billion out of the 300 billion lines of computer code in the whole world.

One seminal handbook from 1971, Gerald M. Weinberg's *The Psychology of Computer Programming*, criticised the sexism of the industry: 'In many projects, women are systematically excluded from management positions, or management positions above the first level,' he wrote. 'To be sure, male managers can offer all sorts of rationalizations for such policies, but such rationalizations accompany any prejudice.' Weinberg concluded: 'The prejudice against women is so common in programming that it merits special attention. Possibly the greatest single action to relieve the shortage of programming and programming management talent would be to start treating women as true equals – if indeed they are only that.'[31] How gallant! I like to think that Grace Hopper, for her part, would have been amused when, in 1969, the Data Processing Management Association presented her with its 'Man of the Year' award.[32]

Hopper eventually won. But initially her power-up idea wasn't adopted: partly because it threatened the status quo, and partly because it came from the wrong kind of person – a woman. (To this day sexism is a powerful force in the computer industry, as confirmed by a 2016 study finding that code written by women was more likely to be approved by other coders than code written by men – but only if their peers didn't know that the code came from women.)[33] Before Grace Hopper's time, extending voting rights to all women was a power-up, too, and the Suffragettes met enormous resistance to this demand. How often might other empowering ideas be rejected simply because they are proposed by the disempowered?

The sexism and parochialism of the early computing industry took a long time to change. We may be years behind where we could have been, technologically, if the work of Hopper and other pioneers had been adopted more quickly in the first place. In other fields, the delay while a new idea has to wait for society to catch up can be positively life-threatening. Many people died needlessly of cancer in the decades after the electronic cigarette had been invented, and before its eventual rediscovery, but the new version may be a revolution in public health.

Similarly, the rejection of atomism for so long may well have delayed the development of life-saving technologies, but at least now we enjoy its benefits in molecular pharmacology and other fields. And in future the embrace of edible insects could, according to its proponents, forestall major famines. Cultural attitudes may change regrettably slowly, but when they do, rethinking can save lives.

Part II: Antithesis

6

Something New under the Sun(s)

Not every idea has been thought of before. Yet even the apparently novel usually has more of the past in it than is often credited.

'The thing that hath been, it is that which shall be; and that which is done is that which shall be done; and there is no new thing under the sun.' – Ecclesiastes

The shoulders of giants

As we have seen, innovation is often driven by the rediscovery and upgrading of old ideas. But it would be wrong to conclude that there is nothing new under the sun. In fact, to do so would be a giant retrograde leap in the history of thinking about the history of thinking. For the idea that *all* new ideas are really old ones was the dominant view of intellectual history before the scientific revolution of the sixteenth and seventeenth centuries. As Geoffrey Chaucer put it in 1380, in his 'Parliament of Fowls', just as new corn comes out of old fields, so 'all this new science' that men learn comes out of old books.[1] The ancients, it was commonly supposed, already knew everything there was to know, but some of their wisdom had been lost over the years. All discovery, therefore, was really rediscovery. Nothing historically unprecedented and completely novel could be thought. And this idea was laid down in ancient texts – so, as it were, confirming itself. As Montaigne wrote in 1580: 'Aristotle says that all human opinions have existed in the past and will do so in the future an infinite number of other times; Plato, that they are to be renewed and come back into being after 36,000 years.'[2]

But sometimes there really is something new under the sun. There was a time, after all, when the mechanical clock, the compass, and the telescope simply did not exist. In our era, epigenetics is a novel science, while also making Lamarckian thinking viable once again. Indeed, the idea that everything was already known in classical times became increasingly untenable from the late sixteenth century, with developments in mathematical astronomy, and the invention of new scientific instruments. This was the age of what the historian of science David Wootton calls the revolt of the mathematicians against the philosophers, most of whom had long assumed that Aristotle was right about everything.[3] It turned out that he hadn't been. Even so, it took a while for the appeal to ancient authority – both classical and scriptural – to die out, even among those who really were proposing astonishing new theories.

Isaac Newton himself was not immune to the pull of ancient authority. Famously, he wrote in a letter: 'If I have seen further it is by standing on the shoulders of giants.' (This image dates back to the French philosopher of the twelfth century, Bernard of Chartres, who compared his contemporaries to puny dwarfs atop the statuesque Ancients.) But, as Wootton convincingly argues, this was merely a display of 'false modesty'.[4] For Newton's theory of gravity really *was* unprecedented: it was a version of no prior idea but something altogether new, and it was greeted as such by his astonished contemporaries.

And yet Newton, having published his groundbreaking work, *Philosophiæ Naturalis Principia Mathematica* ('The Mathematical Principles of Natural Philosophy'), began to doubt that his ideas really could be new. Surely, he thought, the great Greek philosophers must have understood gravity too. We can only speculate as to his motivations, but here might be a paradoxical mixture of authentic modesty (for otherwise Newton would be raising himself above the level of those great philosophers), and egotistical self-affirmation (Newton must be right because these ideas were already held by the ancients, even though none but Newton had yet realised it). But he also really believed his devout thesis: that God had once revealed these perfect truths about the workings of his creation, the universe; that they had then been lost, but partially recuperated and understood by some great sages of antiquity. In 1691, Newton's assistant Fatio de Duillier wrote to the astronomer Christiaan Huygens to explain. Newton, de Duillier reported, had come to believe that Pythagoras, Plato, and other ancients had understood the 'true system of the world' based on the inverse-square law – i.e. gravity – but that they had cloaked their knowledge in 'mystery', expressing it in esoteric terms for adepts only.[5]

Huygens himself wrote back, courteously, to say that he disagreed. The ancients (at least some of them) knew that the Earth went around the Sun and not vice versa, but, he pointed out, they lacked the mathematics to describe the elliptical orbits that it was now known the planets followed. Still Newton persisted. During the 1690s, he made copious notes (now known as the 'classical scholia') towards a second edition of the *Principia* in which he would demonstrate the well-founded antiquity of his own groundbreaking ideas. There were, of course, some moments of brilliant deductive science in ancient philosophy – in his notes Newton mentions specifically, for example, the Atomists' theory of atoms and void. But Newton's argument in these scholia that Pythagoras had discovered the inverse-square law in the harmonies of vibrating strings, and secretly knew that it could also be applied to the movement of celestial bodies, seems much more dubious to modern scholars. A couple of Newton's closest friends continued to humour him in this arcane historical adventure, but the planned new edition never materialised.

Like any innovative thinker, Newton was able to come up with his ideas because his starting point was the state of human knowledge and technology at the time in which he lived – the mathematical tools that already existed before he developed them further, the availability of glass prisms, and so forth. In that sense, no thinker ever works outside of history, and all stand on the shoulders of giants. Nonetheless, the theory of gravity really was novel. As Wootton points out, Newton didn't look for clues in ancient philosophy *before* developing it; he only sought to reconstruct them afterwards.[6] Perhaps because his own discovery was so terrifying in its cosmic majesty, it needed to be tamed by the thought that it had been known all along.

It is the success ever since Newton of the grand story of scientific discovery and technical innovation, however, that has led to a culture today in which the emphasis is too much on novelty and disruption. The ruling idea of innovation is that it must at all costs be original, unprecedented, an abrupt break with the past. Its champions have forgotten how often discovery really *is* rediscovery. This doesn't mean we should return to the seen-it-all-before ambience of pre-modern times, with its eternal cycles of history and opinion and its total deference to ancient philosophers. But it is time for the pendulum to swing back some of the way.

The Gernsback continuum

I am sitting on the sofa in my apartment, wearing a matt rubber headband that uses a set of flexible electrodes to read my brainwaves and relay the data wirelessly to my smartphone. This is an unprecedented technological situation: just a few years ago it would have sounded utterly fantastical. And yet this futuristic gadget also embodies a set of ideas that are much older than Bluetooth.

The headband is a commercially available consumer electroencephalograph (EEG) device called Muse. It comes with premium-styled Apple-like packaging and a very hard sell. Muse will help you 'Do more with your mind', by teaching you how to calm it. Because, the box explains, 'Once your mind is calm, your focus can become clear. Your perception can sharpen. Your ideas can flow more readily and with greater purpose.' That would certainly be welcome. So I adjust the Muse to sit across the middle of my forehead, with the ends of its arms resting behind my ears. Over my earphones I will hear the gentle lapping of waves on a beach, and the occasional gust of wind. I am supposed to concentrate on my breathing, counting each breath. If my mind wanders from this task, the headband will notice a crescendo in brain activity and the wind will get noisier. In a faintly threatening tone, a woman's voice promises: 'Muse WILL sense if you lose focus.' When that happens, I must return 'focus' to my breathing so the wind quietens down again.

There are as yet no peer-reviewed studies supporting Muse's effectiveness. But I do get 'better' at whatever it is measuring, according to the stats on my phone. A few days in and I spend 60 per cent of one three-minute session being 'calm', and only 11 per cent of it 'active'. (The rest is 'neutral'.) Muse HQ probably knew that, because its smartphone app sends your data to its servers, where who knows what NSA-style interpretive operations are performed on it. It could easily be collecting more information than it tells the user about. Perhaps the investor class is excited by Muse's potential user base of exceedingly calm, compliant, and suggestible consumers.

The assumptions behind Muse's approach to mood engineering are, though, questionable. It's useful to be able to quieten the mind on demand, though Muse's use of negative feedback (wind noise) is very different from the meditative traditions that warn against striving towards any particular state. But more importantly, 'focus' as a zombie-like tranquillity is not the kind of focus I – or perhaps you – need. Think hard about a complicated subject wearing Muse and it punishes you with a storm. But that's the kind of brain activity that gets ideas going, while Muse thinks the perfect mental state is monolithic focus on a subject of indivisible simplicity. To aspire always to a placid mindfulness is to be an intellectual hermit. It is the opposite of thinking.

Muse is a dazzlingly novel technology. Yet it belongs to a class of modern devices and apps based on an old tradition in psychology, whose return we might not all be inclined to welcome.

In 1925, the inventor Hugo Gernsback announced in the pages of his magazine *Science and Invention in Pictures* his creation of the Isolator. It was a metal full-face helmet, somewhat like a diving helmet, and it connected by a rubber hose to an oxygen tank. The Isolator was designed to defeat distractions and aid mental focus.

The problem with modern life, Gernsback wrote, was that the ringing of a telephone or doorbell 'is sufficient, in nearly all cases, to stop the flow of thoughts'. Inside the Isolator, though, sounds are muffled. And the small eyeholes prevent you from seeing anything except what is directly in front of you. Gernsback provided a salutary photograph of himself wearing the Isolator while sitting at his desk, looking like one of the Cybermen from *Doctor Who*. 'The author at work in his private study aided by the Isolator,' the caption reads. 'Outside noises being eliminated, the worker can concentrate with ease upon the subject at hand.'[7]

Modern anti-distraction tools such as computer software that disables your internet connection, or word processors that imitate an old-fashioned DOS screen, with nothing but green text on a black background, as well as the brain-measuring Muse headband – these are just the latest versions of what seems an age-old desire for technologically imposed calm. But what do we lose if – unable to impose calm on ourselves – we come to rely on such gadgets?

In addition to the Muse, a wider wave of apps to optimise not only the brain but the whole body smuggles back into the present some long-discarded assumptions in psychology. They were first motivated by the following line of reasoning. No one knows how the mind works, so why not just think of it as a black box, and concentrate instead on the inputs and outputs? The inputs are what we hear and see, and so on; the outputs are our speech and action: our physical behaviour. And so the school of 'behaviourism' in mid-twentieth-century psychology arose, treating human beings as animals that could be mechanistically conditioned just as easily as lab rats.

At an extreme, behaviourism not only supposed that speculation about the mind within was useless; it actually denied the existence of the mind altogether. But behaviourism eventually was supplanted by what is known as the 'cognitive revolution' in psychology. Interventions such as CBT focused on an individual's cognition – the thoughts and the beliefs that behaviourism treated as merely hypothetical constructs

– and saw demonstrable clinical success. The mind *was* important, after all.

Yet the old assumptions of behaviourism are creeping back. Cut to the unveiling in late 2014 of the Apple Watch, with its heart-rate monitor, its altimeter for calculating how many steps you've walked up, and its suite of fitness and no doubt productivity-enhancing apps. It is part of a modern industry that includes other 'wearable' devices such as Fitbit and their associated software, which offer ways to take constant fine-grained measurements of our physiological state – even while we're asleep – and reward us with positive feedback when we improve our running times or lower our resting heart rate over a period of days and weeks. Nothing wrong, of course, with getting fitter and more healthy. But what are we doing when we become entranced by our own data-flow?

A paradigmatic image of behaviourist science is the 'Skinner box', named for one of behaviourism's chief theoreticians, B. F. Skinner. (Formally, it is known as an 'operant conditioning chamber'.) Skinner boxes come in different flavours, but a classic example is a box with an electrified floor and two levers. A rat is trapped in the box and must press the correct lever when a certain light comes on. If the rat gets it right, food is delivered. If the rat presses the wrong lever, it receives a painful electric shock through the booby-trapped floor. The rat soon learns to press the right lever all the time. But if the levers' functions are changed unpredictably by the experimenters, the rat becomes confused, withdrawn, and depressed.

Skinner boxes have been used with success not only on rats but on birds and primates too. So what, after all, are we doing if we sign up to technologically enhanced self-improvement through gadgets and apps? As we manipulate our screens for reassurance and encouragement, obediently click on clickbait, or wince at a painful failure to be better today than we were yesterday, we are treating ourselves, too, as objects to be improved through operant conditioning.[8] We are climbing voluntarily into a giant Skinner box.

Yet there are political and sociological reasons to resist the pitiless transfer of responsibility for well-being on to the individual in this way. And in such cases, it is important to point out that a new idea is a repackaging of a controversial old idea, since that challenges its proponents to defend it explicitly. The Muse brain-sensing headband promises an utterly novel form of technologically enhanced self-mastery. But it is also merely the newest way in which modernity invites us to perform operant conditioning on ourselves, to clean away anxiety and dissatisfaction and become more productive citizen-consumers. Perhaps we will decide, after all, that tech-powered behaviourism is good. But we should know what we are arguing about. The rethinking should take place out in the open.

Mutually assured destruction

As we saw, the US experience in Iraq and Afghanistan – when experts suddenly had to scramble around for old counterinsurgency textbooks because no one had thought about guerrilla warfare for decades – shows that, in the theatre of war, an assumption that new circumstances change everything so that the old rules don't apply is perilous. The Muse headband and other wearable gadgets are novel machines, yet also implicitly revive older arguments in psychology. In war, too, unprecedented new technology can still usually be slotted into a continuity of thinking. The most influential early ideas on the military uses of the aeroplane simply worked by analogy to naval warfare: just as a fleet sought command of the seas, so an air force would aspire to command of the skies.[9] Modern drone warfare is new in that it offers a convenient method of risk-free remote assassination, but it has not yet changed the century-old fact that you cannot win wars – or even compel a determined leader to change course – from the air alone. This fact was recognised very early on in the development of military aviation, even if it continues to slip the mind of gung-ho leaders in our day. On the other hand, it took a visionary in the latter years of the First World War to see how tanks really would revolutionise the conduct of future conflicts.[10]

There is perhaps only one example of a new technology in war apparently rendering earlier thinking completely useless, but it is a powerful example: the atomic bomb. When he first heard of the nuclear strikes on Hiroshima and Nagasaki, the American military historian and theorist Bernard Brodie told his wife: 'Everything that I have written is now obsolete.'[11]

An all-out nuclear war, Brodie rapidly decided, was unthinkable, and he became a leading proponent of 'deterrence' strategy: developing a nuclear arsenal capable of wiping out the enemy simply to ensure that the enemy would never attack. In 1946 he published his seminal book, *The Absolute Weapon: Atomic Power and World Order*, in which he famously wrote: 'Thus far the chief purpose of our military establishment has been to win wars. From now on its chief purpose must be to avert them. It can have almost no other useful purpose.'[12] A new generation of civilian strategists applying economics and game theory to the possibility of nuclear confrontation congregated at the RAND Corporation and then the government throughout the 1950s.

Game theory is the mathematical analysis of co-operation and conflict. It was the joint invention of the great mathematician John von Neumann – who, as we saw, worked with Grace Hopper during the war – and the economist Oskar Morgenstern. Von Neumann, in fact, argued forcefully to President Eisenhower that the nuclear stand-off between the US and the USSR was a game of Prisoner's Dilemma in which the only rational choice for either player was to launch a massive first strike to wipe out the enemy. So, he said, the US should launch an all-out nuclear attack immediately.[13] We may be grateful that von Neumann's influence on his times was not total. Cooler heads prevailed, including that of Thomas Schelling, who in 2005 was awarded the Nobel Prize in Economics for 'having enhanced our understanding of conflict and cooperation through game-theory analysis' during those mid-century crisis years. Schelling in fact became the dominant theorist of ways to use nuclear weapons without actually detonating their warheads.

Nuclear weapons, Schelling argued, were to be thought of primarily as an instrument of political coercion – of deterrence (to stop the enemy attacking), and positively, in the jargon of the time, of compellence: to make the enemy do something else. Schelling pioneered detailed conceptual analyses of nuclear 'escalation' that might stop short of all-out war, and strategies of direct and indirect communication between nervy governments that could avoid Armageddon even if some steps on the escalation ladder had already been taken. All this analysis was – luckily for millions – never put into practical effect. Yet that was precisely its practical effect: by breaking down possible steps on the ladder, and emphasising tension, bargaining, and communication between adversaries, Schelling's analysis enabled the military and politicians to avoid making the automatic jump in their thinking from conventional engagement with armies and air forces to a single massive nuclear conflagration.

During the Berlin crisis of 1961, for example, the Soviets threatened to absorb West Berlin, an annoying outpost of the West that lay deep inside East Germany. The Americans could not defend it by conventional means, so any attempt to prevent the USSR's plans would risk nuclear war. At this time, John F. Kennedy was deeply impressed by the message of a paper on nuclear strategy that his advisors passed to him. It was by Thomas Schelling, and its message was: 'We should plan for a war of nerve, of demonstration and of bargaining, not of tactical target destruction.' Schelling helped to set up a war game in Washington that took participants through many rounds of bargaining over the Berlin question. The game taught its players that it was surprisingly difficult to get a war started; but once one was started, nothing short of total annihilation would be effective.[14] In the event, a war of nerves was indeed thought preferable to a war of nukes. A compromise result kept the peace. The USSR erected the Berlin Wall.

So the recent intellectual invention of game theory, used by John von Neumann to recommend the immediate killing of millions of Russians, was used more successfully by Schelling to avoid any nuclear launch at all. Or was it? Schelling certainly referred to game theory in his writings, but that isn't where he thought the real wisdom lay.

In a later interview he explained what the truly relevant questions were when thinking about strategy in the nuclear age. 'My interest has always been in the ways that people can make believable threats,' he said. 'What are the institutions that allow it, what legal arrangements allow it, how does it depend on the capacity for reputation? How does it depend on the use of agents? How in the world do nations, people, or organizations go about committing themselves to threats and promises in bargaining positions?'[15] Schelling said that the ideas he had found most influential on his own work came, in fact, from his analysis of salesmanship and ancient Greek history. His work describes, for example, strategies of negotiation for agreeing on the price of a house or car, and how deterrence operates in ordinary situations such as 'the threat to bump a car that does not yield the right of way or to call a costly strike if the wage rate is not raised a few cents'.[16] Meanwhile he draws important lessons on strategy from the Greek commander Xenophon, who 'when threatened by an attack he had not sought [. . .] placed his Greeks with their backs against an impassable ravine'.[17] In this way Xenophon made sure he could not back down, in order to make his threat to fight more credible to his enemy. And this strategy of what Schelling called 'precommitment' was one he recommended to leaders thinking about nuclear deterrence. So it was a man consciously rethinking the past who helped to preserve the present.

So far, at least. The prospect of 'mutually assured destruction', aptly acronymised as MAD, was what kept the US and the USSR from launching nuclear attacks throughout the rest of the cold war. The idea that rationally calculating actors must realise that such a war could not be fought was vividly popularised in the film *WarGames* (1983), in which a supercomputer learns that global thermonuclear war is a game in which everyone loses. But what if the rational calculation changes?

In January 2016, a magnitude 5.1 tremor was detected in Punggye-ri, North Korea. The North Korean government claimed that it had successfully tested a small hydrogen bomb for the first time, after previous tests of simpler atomic bombs. A handwritten letter from Kim

Jong-un was broadcast on state television, which read: 'Let's begin the year of 2016 – a glorious and victorious year when the historic seventh conference of the Workers' Party of Korea will be held – with the thrilling sound of the first hydrogen bomb explosion, so that the whole world will look up to our socialist, nuclear-armed republic of Juche and the great Workers' Party of Korea!'[18]

Unthrilled observers, including South Korean and Swedish officials, were sceptical of the H-bomb claim, pointing out the explosion's relatively low implied yield of six kilotons, less than half the power of the atomic bomb dropped on Hiroshima. A true H-bomb, using nuclear fusion – also known as a thermonuclear device – would normally be thousands of times more destructive. Nonetheless, what North Korea tested was *some* kind of nuclear weapon, and big enough to wipe out a medium-sized city. Most notably, commentators once again began reaching for the jargon of international nuclear confrontation that had been more or less gathering dust for two and a half decades. Britain's former ambassador to North Korea, John Everard, pointed out that an elaborate system of underground tunnels lay beneath Pyongyang, which might persuade the regime that 'they have survivability' – in other words, that they could survive a nuclear attack.[19] In which case, deterrence might no longer work.

In the mean time, the United States had been developing a new nuclear bomb, the B61-12, designed to target underground installations more precisely. Because the B61-12 was planned to have a maximum 50-kiloton explosive yield, proponents celebrated the reduction in radioactive fallout and 'collateral damage' the weapon would cause. Critics in the military and nuclear establishment, however, argued that this attitude made it more likely that such weapons would actually be used, risking escalation into all-out nuclear conflict.[20] 'What going smaller does,' acknowledged one of the new weapon's backers, General James E. Cartwright, 'is to make the weapon more thinkable.'[21] And that is precisely why the 'tactical' use of smaller nuclear weapons was avoided throughout the cold war.[22] Maybe the old ideas about nuclear strategy, too, would have to be revisited, relearned, and upgraded in the twenty-first century.

Many worlds

There is always something new under the sun, it seems, in physics – an intellectual engine for the production of novel, exotic fauna such as strings, branes, and black holes. The field of modern cosmology, in particular, is full of epically weird ideas. And these days, a lot of people whose job it is to think about why the universe is the way it is aren't satisfied with the problems posed by just one universe. They have decided there are lots of universes. Maybe infinitely many. Welcome to the multiverse.

It is the kind of crazy sci-fi notion we are pleased to have experts considering. And it might also be a strange kind of consolation. After Zayn Malik left the band One Direction in 2015, Stephen Hawking offered some advice to fans. 'My advice to any heartbroken young girl,' he said, 'is to pay close attention to the study of theoretical physics. One day there may well be proof of multiple universes . . . somewhere outside our own universe lies another, different universe. And in that universe, Zayn is still in One Direction. This girl might like to know that in another possible universe, she and Zayn are happily married.'[23] The potential existence of multiple universes seems like a poster child for the brain-stretching character of ideas in twenty-first-century science. But some thinkers found themselves driven to it thousands of years ago. The multiverse, too, is older than it seems.

The young cosmologist Andrew Pontzen greets me in the front court of University College London and leads me to a corner table in the sunny common room to discuss the origins of everything. He has sideburns and hip, technical spectacles, and a much more historically sympathetic attitude to bygone ideas than some of the big beasts of science popularisation. Pontzen had said almost casually on a radio programme that the steady-state theory in cosmology had, in a way, returned. Wait, really? But everyone knows now that the steady-state picture of the universe was *just wrong*.

This idea was originally proposed in the mid-twentieth century by the great British astronomer Fred Hoyle. He pioneered the extraordinary discovery of how heavy elements are synthesised inside stars, but for mysterious reasons did not share in the Nobel Prize awarded for the work. Hoyle did not believe in the Big Bang – indeed, it was he who derisively coined that name for it. At the time, everyone already knew that the universe was expanding. Well, if you just run the film backwards into the past, so that it looks like the universe is contracting rather than expanding, what does that imply? To many people it implied that at some point very long ago the universe must have started out in a tiny point that was extremely small and extremely hot – the moment of the Big Bang. That idea is what Hoyle described as 'the assumption [. . .] that the Universe started its life a finite time ago in a huge explosion' – a theory originally proposed by the Catholic priest Georges Lemaître.[24] Hoyle couldn't believe it, and nor could many others; in fact, much of the early resistance to the Big Bang was motivated by a suspicion that it was just religion in scientific garb. (The idea was welcomed by Pope Pius XII, who declared it evidence for the 'primordial *Fiat lux*' – let there be light – 'when, along with matter, there burst forth from nothing a sea of light and radiation, while the particles of chemical elements split and formed into millions of galaxies'.)[25]

Instead Hoyle and his supporters preferred a 'steady-state' theory. Yes, the universe was expanding, this idea said, but new stuff was constantly being created to fill in the gaps. 'So two galaxies fly apart,' Andrew Pontzen explains, 'but in the mean time another galaxy gets born in the gap that's left, and so overall the universe looks the same from one time to the next.' And things had always been that way, steady and eternal. There was no beginning to the universe.

What killed that idea was the discovery, in the 1960s, of the Cosmic Microwave Background. It is a faint radio noise that pervades space, which we interpret as the faint afterglow or echo of the Big Bang itself. (Detune an old analogue television: some of the static you see on screen is actually caused by the CMB, in other words by the Big Bang.) On the basis of this compelling finding, most scientists abandoned the steady-state theory. One, Hermann Bondi, called the CMB a highly convincing 'fossil' of the Big Bang.[26] Hoyle, on the other hand, clung to his steady-state theory until his death in 2001, to sadness or ridicule from his colleagues.

The Cosmic Microwave Background is the biggest-ever smoking gun. (It doesn't get us all the way back to the Big Bang, because it originates 300,000 years later, but most accept it as very compelling evidence that the Bang went . . . bang.) And yet something very like what Hoyle proposed instead of the Big Bang, an eternally existing 'steady-state' universe with no definite beginning, has returned. Curiously, it seems the two ideas can coexist. Not, perhaps, in the same universe – but at least in the same multiverse.

Back to realities

Our word 'universe' comes from the Latin *universum*, meaning 'the whole' or 'the sum of all existing things'. And yet modern scientists increasingly suspect that our universe is just one of many, so it's not the sum of all existing things after all. This idea of a multiverse – multiple universes, each as mind-bendingly vast as the one we live inside – has an exotically modern feel. (The 'inflationary multiverse' model in modern cosmology has been around only since the 1980s.)[27] But it too was present at the very birth of natural philosophy, in ancient Greece.

Democritus, our laughing philosopher, didn't just think small, about bread and atoms. He thought big. He wondered: how many universes are there? That seems a strange question. Why isn't just the one enough? Well, the conceptual trouble starts if you try to imagine what is outside the universe. (If there's something outside it, then the 'universe' should be defined as the universe plus that empty space. But then what's outside that?) More trouble comes if you ask what came before the universe. (Another universe? Aha. Or nothing? If so, where did the universe come from?)

Democritus and his Atomist friends, worried by the first question – what is outside the universe? – decided that there had to be a multitude of universes, or *kosmoi*, zooming around and sometimes colliding destructively in infinite space. Vehemently opposed to the crazy theories of the Atomists were the Stoics. As we have seen, their ideas about how to maintain mental equilibrium reappeared in modern psychology two thousand years later. But the Stoics were also worried by this second question – what came before the universe? And so they concluded that there must be an eternal cycle of births and deaths of the universe.

Such interesting but apparently undecidable cosmological theories never became mainstream even in Greek philosophy. Plato hated the idea of multiple universes. The cosmos had to be singular, he declared, because the ideal Form of the perfect living creature was singular. From then on, one universe was enough for most people, even well into the twentieth century. The notion of a multiplicity of universes seemed reckless and unprovable to most scientists. Until quite recently.

Eternal inflation

The Big Bang theory was not without its own mysteries. It did not explain why the universe began in the particular state that it did. And it proved particularly difficult to explain the present universe's apparent smoothness. 'If we look through space,' Andrew Pontzen says, 'we find that things are, on the whole, very uniformly spread out if you look on large enough scales.' (In technical terms, the universe appears to be 'homogeneous', smoothly spread out, and 'isotropic', favouring no particular direction.) Many different ideas were tried to explain this smoothness, and the best candidate came to be known as inflation. 'That's the idea that some exotic physics really pushes the universe apart at a very rapid rate early on,' Pontzen says, which guarantees the homogeneity we observe today. Inflation, he notes wryly, 'seems rather made-up'. 'It's not based on stuff that we can go and test in a laboratory. On the other hand it made a whole load of predictions, and one of the predictions was for the exact types of patterns you should see in the CMB, which was subsequently confirmed.'

So far, so good. But the problem with this kind of powerful, exotic inflation, as it turns out, is that it has a tendency to run away with itself. 'People began to realise that stopping inflation is really hard,' Pontzen says. 'If you have this substance that is trying to push the universe apart, getting rid of that substance and turning it into the normal particles and stuff that we know and love is a difficult process. And so people realised that actually for a very wide variety of scenarios, if inflation is correct, then it must be *eternal* – it must carry on for ever, and it's only in very lucky patches that it ever ends.'

Let us pause for a moment here to give thanks that we are living in a lucky patch. The remainder of creation, beyond what we can directly observe, is continuing to inflate. And it turns out that mathematically, this theory is '*extremely* similar to Hoyle's steady-state picture. Just that now you're doing it with universes instead of galaxies.' So, just as Hoyle thought the eternal universe was expanding with new galaxies appearing to fill in the gaps, the new theory says that the eternal *multiverse* is expanding with new universes appearing to fill in the gaps.

Pleasingly, then, we have an example of rethink inside rethink: Fred Hoyle's moribund steady-state theory, which opposed the Big Bang idea that he mocked, turns out to work surprisingly well as a description of the once-mocked idea of a multiverse. In a way, Hoyle's idea just needed to be repurposed to work at an even vaster scale.

Unfalse

But how could we ever know the multiverse actually exists? We can only ever see what's in our own universe. In that case, any multiverse theory would seem to be unprovable – or rather, as one tradition in the philosophy of science puts it, unfalsifiable. If there is no conceivable counter-evidence that could turn up and *disprove* a theory, say the proponents of falsifiability, then it is not a respectable theory at all.

Well, so much the worse for falsifiability, say some scientists. Cutely, the physicist Frank Wilczek has taken inspiration from the comedian

Stephen Colbert's coinage 'truthiness' and suggested that, instead of falsifiable, a good theory should be 'truthifiable'. If the theory suggests further testable experiments and ideas, then it can be useful even if it is wrong. 'A truthifiable theory might make mistakes,' Wilczek writes, 'but if it's a good theory they're mistakes you can build on.'[28] In this way truthifiability is rather like Jochen Runde's reversal of the Baconian scientific method. Instead of trying to disprove your hypothesis, you look for evidence that might support it – and in doing so you might discover useful new things, even if the hypothesis itself is inaccurate.

The prominent cosmologist Sean Carroll, for his part, has argued that now that science is able to probe fundamental reality so deeply, the traditional requirement for any scientific theory to be empirically testable ought to be weakened if not abandoned. After all, the multiverse doesn't care about our philosophical scruples. If the cosmos is like that, then it's like that, regardless of whether we'll ever be able to know for sure. It's time, in other words, to rethink our conceptual picture of how science should proceed. Experiments are nice, but sometimes you just can't do them; and that doesn't mean you're not still doing science. Others disagree strongly: late in 2014, George Ellis (a mathematician) and Joe Silk (a physicist) published a letter in *Nature* deploring the untestable claims of multiverses and string theory and insisting that 'post-empirical science is an oxymoron'. 'The scientific method is at stake,' they warned.[29] The debate rages on.

'Sean Carroll must be right on some level,' Andrew Pontzen muses. 'If the universe is that way, it is that way, and it's a real shame if we can't test it – but that doesn't mean it's wrong.' Happily, though, there might be a way out of this particular disagreement. 'To my mind even more excitingly, maybe it *is* possible to test this,' Pontzen says. 'Because if you have these universes popping up at different points inside the multiverse, then occasionally two universes will actually bump into each other.' (Recall the crashing *kosmoi* of the Atomists.) And if that has happened within the lifetime of our universe, there ought to be a giveaway relic in the form of a faint pattern in the Cosmic Microwave Background – 'a disturbed ring' resulting from the epic collision of two universe-bubbles.

This potential test distinguishes the modern (and ancient) multiverse concept in cosmology from a related and mind-boggling idea in philosophy, known as modal realism. Proposed in the 1980s by the British philosopher David Lewis, modal realism holds that any possible world we can imagine – where a 'world' is the entire universe, and 'possible' just involves no logical contradiction – is a world that really exists. The thesis is put with shocking simplicity: '[A]bsolutely *every* way that a world could possibly be is a way that some world *is*.'[30]

In other words, a world where everything is the same as this one except that from now on this book consists entirely of pictures of cats – that world is not merely hypothetical; it really exists, just as much as this world does. In case anyone doubts that Lewis is really saying what he seems to be saying, he also insists that there are infinitely many donkeys. 'Besides the donkeys among our worldmates,' he writes happily, 'there are countless other donkeys, spread over countless worlds.'[31]

I could go on, but you get the idea. (In another universe, I *am* going on.) There are infinitely many worlds. But you'll never catch them crashing into one another, because, Lewis argues, they are perfectly isolated. 'There are no spatiotemporal relations at all between things that belong to different worlds. Nor does anything that happens at one world cause anything to happen at another.'[32] So: no colliding *kosmoi*; no disturbed rings in the CMB as a giveaway. It seems that these alternative universes can never in principle let slip evidence of their existence. So why believe in them at all? Just because, Lewis argues, the hypothesis is useful. Talk of possible worlds, he points out, had enabled a lot of progress to be made in philosophy during the 1970s. (It advanced the analysis of modal logic – discussions of the possibility or necessity of certain truths.) '[T]he hypothesis is serviceable,' Lewis writes, 'and that is a reason to think that it is true.'[33] This is quite startling. Lewis does not argue that an idea's usefulness is *sufficient* reason to think it is true; it is just 'a reason'. Yet some hypotheses may be useful even if they are false. As we'll see later, the usefulness of an idea may be good reason to go beyond judgements of truth or falsity altogether.

What modern cosmologists can get on with, in the mean time, is the relatively more parochial task of trying to figure out whether our universe is one of many or not. And that's just what they are doing. 'We're trying to do calculations,' Pontzen says, of 'how likely is it for two of these universes to collide, because we just don't really know whether that would be very rare or quite common.'

Quite: neither do I. But look: the multiverse, which began as purely logical speculation, is now being rigorously mathematically modelled. It's a hi-tech rethink of a grandiose old idea. And this is what is truly new under the numberless suns of the multiverse. The mathematics of today was not available to the ancients, so they could not have such a developed theory of the multiverse as modern cosmologists. And yet the fundamental questions that puzzled the Atomists and the Stoics – what could be outside the universe? what could have been there before it existed? – have lost none of their motivational power two millennia later.

Back to Plato

Rationalism is the very old idea that we can apprehend fundamental aspects of reality through pure thought, just as Democritus did when he understood the necessity of atoms. Rationalism is usually rhetorically opposed to empiricism, the reliance on experience or experiment. In our day of evidence-based everything and Big Data, it might seem as though empiricism is all, and rationalism the superstition of a bygone age. 'Empirical' is almost synonymous with reliable, or true. 'Armchair thinking', on the other hand, is usually a term of abuse: that alone won't get you anywhere. Except that, very often, it does.

For one thing, mathematics – an elaboration of pure thought – has a very mysterious relationship with reality. Its findings, about the distribution of prime numbers or the truth of Fermat's last theorem, seem to be utterly correct and independent of whatever we discover about the world around us. What's *really* strange is that mathematics is also the language in which we can write down physical 'laws' of nature with which to make predictions about the future, and thereby make pretty much the whole of science and engineering possible. (That is so odd that there is a famous 1960 paper by the Nobel-Prize-winning physicist Eugene Wigner entitled 'The Unreasonable Effectiveness of Mathematics in the Natural Sciences'.) And, as Frank Wilczek demonstrates, a taste for beauty in mathematics has, surprisingly often, been an accurate guide to what turns out to be the most useful idea. Often, theories constructed according to a preference for mathematical symmetry (the very specific definition of 'beauty' that Wilczek is using) long precede the moment when it is finally possible to validate them. James Clerk Maxwell, for example, predicted 'new colors of light, invisible to our eyes', which no one had ever detected – until Hertz produced radio waves and proved him right; Paul Dirac predicted the existence of antiparticles; and so on.[34] It has seemed pretty astonishing to many scientists throughout history, indeed, that the workings of the universe are intelligible to human reason at all, which has led many of them, including Newton, to assert the existence of a rational Designer – not necessarily an interventionist God, but at least a being (in what is known as the deist view) who set up the whole thing on rational principles in the first place, so that rational beings such as us could understand them. Even if one rejects the notion of a creator, the assumption that nature will go on being comprehensible to our powers of reason is simply an article of faith.

Armchair thinking, too, is what produced the theory of cosmic inflation, which might seem 'rather made-up', as Andrew Pontzen puts it, but then made very accurate predictions. And one of the most spectacular successes of armchair thinking in history was Einsteinian relativity. As Pontzen points out: 'Einstein's formulation of gravity came out of pretty much pure thought, and then he went and said: "Oh look, it explains [the orbit of] Mercury, and deflection of light, and all of these things."'

On the other hand, the idea of dark matter – an exotic type of matter that we can't see, but which is postulated to make gravitational equations add up – arose purely out of considering difficulties in available evidence: 'You had to explain why the galaxies behave the way they do,'

Pontzen says, 'and it seemed to work for that, and then the theorists sort of took over and said: "Ah, okay, so maybe it ties in with *here*, and maybe it's created like *this*, and the *following* would happen", and you make some new predictions and you go and test them again.' In this way, rather than being mortally inimical philosophies, rationalism and empiricism work beautifully as a tag team. The armchair thinking of millennia ago (in the case of atoms and the multiverse) stands up surprisingly well at the frontier of science today: a faith-based rationalism that can peacefully coexist with a careful attention to evidence and data.

This is particularly true of modern computer-mediated science that involves extensive mathematical modelling and simulation, which is now often indispensable for conducting experiments. Biomathematics, for example, is a huge and growing field that uses models to simulate the behaviour of cells in much more detail than is otherwise possible. And in modern high-energy physics, it can be difficult to draw a clear line between theory and experiment at all. Experiments in physics have changed radically over the past hundred years. We might visualise scientists of the late nineteenth century experimenting in laboratories, producing electromagnetic waves with tabletop devices, pounding sheets of gold with gamma rays, measuring light through great brass telescopes. Now high-energy physicists need a 27-kilometre underground ring of superconducting magnets – the Large Hadron Collider at CERN. In this remarkable facility, the measuring apparatus itself is a series of extremely complex devices, engineered according to many theories. What it measures is interpreted according to computer models of how the apparatus should behave. In turn, there are computer models of how the subatomic particles themselves are behaving, and computer models of how the vast sets of data generated by the computers should be interpreted. The very idea of a scientific 'observation', then, is much more complicated than it was in the age of tabletop physics experiments: it arises from the interaction of many interdependent models and theories. So, the philosopher of science Margaret Morrison argues, perhaps 'we need to rethink the way we think about experimental measurement' altogether.[35]

When you are using mathematics to model aspects of reality in a simulation, and then using the results of that simulation to tell you what is going on in an experiment, then Big Data and pure reason exist in an unprecedented symbiotic fusion. In our time, the sharp line between experiment and theory has been rethought out of existence. Just as the multiverse is both cutting-edge cosmology and ancient logical necessity, the Large Hadron Collider is both a tremendous gadget of modern industrial-scientific inquiry, and a machine of pure Platonism.

Yes, sometimes there is something new under the sun: telescopes, computers, online dating. But the most novel-seeming theories and technologies of the present may have unexpected ancestors. Wearable tech is a great wave of Silicon Valley innovation, yet it also implicitly revives the dubious assumptions of behaviourism. Nuclear weapons seemed to render all previous military theory obsolete overnight, yet the psychology of bargaining and threat that informed the new thinking on nuclear strategy would have been familiar to the Greek general Xenophon. The idea that reality is permeated by the fizzing aftershock of the Big Bang – the Cosmic Microwave Background – really was novel, like Newton's theory of gravity. Previous generations never suspected anything like it. And yet, slowly but surely, the fallout from this idea led scientists to a theory of multiple universes that – in its general outlines – would not surprise the philosophers of ancient Greece. It no longer makes any sense to argue, as people did before the scientific revolution, that all discovery is rediscovery. But the shiniest new ideas can still have surprisingly deep ancestral roots.

7

The Jury's Still Out

Some ideas keep coming back even though it might never be possible to confirm them.

'I see bodies that think and feel. I mean bodies: that is, men and animals that I do not see, do not feel, do not know, and cannot know, to be other than bodies. Therefore I say: matter can think and feel; it does think and feel.' – Giacomo Leopardi

Talk to the glove

How do you arrange your sock drawer? Do you ball one sock inside the other and just chuck the pairs in? Or do you neatly fold each pair of socks and stand them up vertically? If you do the latter, you're probably a convert to the KonMari Method, a system for decluttering and tidying the home that originated in Japan and rapidly conquered the rest of the world. It's named after its inventor, the tidying theorist Marie Kondo. Her book explaining the rules, *The Life-Changing Magic of Tidying*, has sold three million copies worldwide, and the idea is endorsed by Gwyneth Paltrow's lifestyle empire Goop, among many others. But it's not just a way of folding your socks and throwing away things that don't 'spark joy' when you touch them. The KonMari method is also a philosophical system.

When you read *The Life-Changing Magic of Tidying*, you can see why it has become a bestseller. The book is extremely charming, witty, and optimistic. But it also wants you to look at the objects around you in a different way. Why, for instance, should you fold your socks rather than balling them? You should do it for the socks' own sake. 'The socks and tights stored in your drawer are essentially on holiday,' Kondo explains brightly. 'They take a brutal beating in their daily work, trapped between your foot and your shoe, enduring pressure and friction to protect your precious feet. The time they spend in your drawer is their only chance to rest. But if they are folded over, balled up or tied, they are always in a state of tension, their fabric stretched and their elastic pulled. They roll about and bump into each other every time the drawer is opened and closed.'[1] Doing things the right way, by contrast, will make your tights and socks 'much happier', and they'll breathe 'a sigh of relief'.[2]

It's almost as if the socks and tights have feelings, right? That's exactly the point. Kondo even wants you to talk to your clothes as you fold them correctly. 'When we fold,' she explains, 'we should put our heart into it, thanking our clothes for protecting our bodies.'[3] When you fold something the right way, 'It's like a sudden revelation – *so this is how you always wanted to be folded!* – a special moment in which your mind and the piece of clothing connect.'[4] And even if you are throwing away an article of clothing, Kondo advises, you should engage in the right sort of dialogue. Don't feel sad or guilty. Instead, say out loud to the doomed garment: 'Thank you for giving me joy when I bought you', or 'Thank you for teaching me what doesn't suit me.'[5] Then you can let it go with a light heart.

The same principle applies when you are embarking on the larger journey of tidying your entire home. Books, for example, should all go on the floor and then get moved around. Not just for your sake, but for theirs. 'Just like the gentle shake we use to wake someone up,' Kondo explains, 'we can stimulate our belongings by physically moving them, exposing them to fresh air and making them "conscious".'[6] Well, of course, our belongings aren't really 'conscious'. Or are they?

In such moments, Marie Kondo is playfully invoking the Japanese tradition of animism, according to which inanimate objects are imbued with spirits or souls. That might sound like outdated religious superstition, yet it's an idea that has recurred again and again in many systems of thought around the world. And something very like it is making a comeback right now in Western philosophy, after millennia of obscurity.

Roll over Beethoven

To return to old ideas is a very common strategy in literature and the arts. While some of his contemporaries pressed on into a world of musical atonality, Igor Stravinsky looked backwards for inspiration and developed a neoclassical style. In our time, Jonathan Franzen abandoned the postmodernism of his early books and made a conscious return to the big Dickensian novel about people and relationships. That brought him public acclaim for works such as *The Corrections* and *Freedom*, but it did not mean that everyone agreed that this was now the only way it was possible to write novels. Other writers, such as Ben Marcus, continued to plough more experimental furrows, to the continued delight of readers who preferred them.

Neither did the novels in Franzen's new Dickensian style, of course, take place in nineteenth-century London. They addressed twenty-first-century American life. A serious artist can never do things exactly the way they were done before. The modernist poets of the early twentieth century revived near-forgotten poetic forms from previous centuries – as with William Empson's devastating sestinas and villanelles – but they did not use those forms for expressions of Renaissance courtly love. And a modern composer could not write a symphony in the style of Beethoven without it being dismissed as mere pastiche. Artists who take something from the past still have to make it somehow new. But when they do, they are not proving anything. No one else has to agree.

In general, artistic styles come and go in a cyclical way, according to the vagaries of fashion and the temperaments of individual artists. Everyone is always rethinking tradition, but no one wins. Ideas in art are never refuted or triumphantly vindicated. The singular execution is all. So the life of ideas in art does not exhibit quite the same dynamics of rediscovery and rethinking that this book is exploring in science, war, and technology.

At first glance the arena of philosophy might look similar. Arguments and ideas recur throughout its history, but there is rarely a moment when everyone is definitively persuaded of the truth of one position or another. There are rival schools, and there are fashions, just as there are fashions in painting or in architecture. And yet there is an important difference. For philosophers make claims about *how things really are.* It may be difficult or impossible to tell if they are right – just as it might be impossible to tell whether we live in a multiverse. Yet philosophical ideas do in some sense have to answer to reality, if only at the end of time.

The ideas we'll meet in this chapter are claims about how the world really is, but they are not as compellingly supported as the theories of physics that underpin aviation and smartphones. Ideas can return and gain significant traction again, even if the jury is still out on the question of their ultimate truth. For their advocates, there are good reasons why they suddenly seem viable or even necessary again after centuries of neglect. These ideas may seem wildly improbable. But improbable is not impossible. They seem to fly in the face of common sense. But then one job of rethinking is to render our common sense uncommon again. So after folding my socks gratefully one morning, I went to see one of the leading modern proponents of such an idea.

Why should the light go on?

The philosopher Galen Strawson smiles under a mane of blond-grey hair as he opens the front door of his light-filled studio in north London. At the back, beyond the piano and sofas strewn with books, Apple laptops, and a classical guitar, is a massive wall of books. 'It's in reverse historical order,' Strawson explains. 'Because I couldn't bear to have Descartes out of reach. So Wittgenstein's up there instead.'

Strawson busies himself making coffee with an inscrutable steel machine of whose workings he appears unsure. ('Oh, have I blown it?' he exclaims at one point, then laughs. 'Sorry. I'm behaving like a classic philosopher, I suppose.') There is amused gossip about a spat in the letters pages of the *Times Literary Supplement* that has followed a recent article of his, and then we sit down and he begins to walk me through the reasoning that, in his view, compels us to accept a quite extraordinary old theory.

'*Astonishing! Everything is intelligent!*'

So runs a fragment attributed to the Greek mathematician and philosopher Pythagoras, two and a half thousand years ago. *Everything* is intelligent. Really? Not just you, but this coffee mug, that rock, the ocean? Marie Kondo's socks?

Yes, that's what he meant. This is a form of the view that at length came to be known as 'panpsychism'. (The modern word is recorded in English from 1879, coined from the Greek for 'mind everywhere', but the idea long predates it.) Mind, it says, must be a fundamental property of *all* matter – of all physical stuff.

It sounds bizarre from our perspective, because we think we know quite a lot about physical stuff. And it is precisely because we think we know quite a lot about physical stuff that there has been something called a 'mind–body problem' for the last few centuries. How does your conscious experience right now, reading this page, arise from the electrochemical interactions of bits of dumb matter in your brain? How do we get from atoms to first-person experience?

Despite what one might assume from popular accounts of neuroscience, the answer is that no one has the faintest idea. As Strawson puts it: you can take matter and 'put it together in this really complex way with this amazing electrochemistry' in the brain, 'but then you still have to cross that great gap, the explanatory gap. Why should these little bits, placed in certain arrangements, whizzing round in certain ways – why should the *light* go on? Why should the light of consciousness go on, if it wasn't there before?'

The answer, for Strawson, is that it *was* there before: experience is already there, deep down, as a fundamental aspect of all matter. This doesn't so much solve the mind–body problem as cut its Gordian knot. If you accept panpsychism, there simply isn't a mind–body problem.

You might think that accepting this astonishing thesis is rather a high price to pay for dissolving the puzzle. And the mainstream consensus throughout most of philosophical history, indeed, has been exactly that. For most of the past two millennia the idea has been considered a mere trick, or an admission of defeat. One can trace a brotherhood of panpsychists through the ages, but the view was always rejected by the mainstream of thinking. Indeed, it was ridiculed as soon as it got started. The atomist Lucretius satirised a panpsychist position with glee:

> If we cannot explain the ability of each animate being to experience sensation without attributing sensation to its constituent elements, what are we to say of the special atoms that compose the human race? Doubtless they shake and tremble with uncontrollable laughter and sprinkle their faces with dewy tears; doubtless they are qualified to discourse at length on the structure of compounds, and even investigate the nature of their own constituent atoms.[7]

As we shall see, that is not the kind of thing a smart modern panpsychist thinks, but it set the tone. Panpsychism was considered disreputable at best, and more usually laughable or infantile, until just a few decades ago – when Galen Strawson, and then a handful of other philosophers, began to take it seriously again.

Choice points

One weirdly popular response to the mind–body problem these days is to say that the mind just doesn't exist. These mind-deniers, sometimes known as 'eliminativists', won't detain us long here, though some of them, like Daniel Dennett, are justly celebrated for other aspects of their work. The way they deny the mind is by saying that, while we think we have conscious experience, this is just an 'illusion' produced by our brain.

This claim, however, is straightforwardly demolished by elementary logic. The experience of an illusion is itself, necessarily, an experience. Something that is not conscious can experience nothing – and that includes illusions. So the very fact – accepted by the experience-deniers – that we *seem* to have conscious experience just guarantees that we *are* having conscious experience. That is what 'seeming' *is*. Or, as Strawson puts it: 'The phenomenon of there seeming to be experience – the phenomenon we're supposing to be an illusion – can't exist unless there really is experience.'[8] It is tempting to say that if we can know anything at all, we can know that we are conscious. (That is the starting point, for example, of Descartes's *cogito ergo sum*. But Descartes was no panpsychist: he argued that mind was a different kind of substance from the rest of matter, and interacted with the body through the pineal gland in the brain. In positing two fundamentally different kinds of stuff, he was a 'dualist'.)

So we know we are conscious, and our consciousness cannot be an illusion because in that case we couldn't experience it. Even so, some people continue to deny the existence of mind at all. 'There are seven and a half billion people who believe in consciousness,' Galen Strawson says, 'and there are maybe two or three hundred philosophers who don't.' In print, he has impishly called the denial of mind 'the silliest view that has ever been held in the whole history of the human race'.[9]

What drives thinkers to such an apparently ridiculous view? Well, most people in our age want to subscribe to 'physicalism' (also known as 'materialism'), according to which everything can in principle be explained fundamentally by reference to the theories of physics, which describe how matter behaves. (This doesn't depend on our having perfect knowledge of the true theories of physics right now. We just need to imagine some future ideal theories that unify all the observable phenomena such as gravity and quantum physics.) Physicalism says that there is no non-physical *other* stuff, no soul-dust. Cartesian dualism – according to which there are basically two substances, physical stuff and mind stuff – is anathema. But, as noted, we have no idea how the physical stuff in the brain gives rise to the mind. So it might seem easier just to say well, mind is an illusion. There's nothing but the dumb atoms.

But that's not the only option. Strawson presents his own argument as a series of 'choice points'. First, most people want to be physicalists. But they also want to accept the reality of consciousness. And if you have made those two choices so far, you are 'on the path' with Strawson. 'Let's draw it!' he says. He puts a blank sheet of A4 paper on the coffee table and sketches a branching diagram of the possibilities. 'One wants to be a physicalist,' he says, 'but one wants to be a realist about consciousness, so of course it follows from those two things already that *consciousness is wholly physical.* At that point you face the question of emergence, radical emergence.'

Emergence is a fashionable as well as useful concept in the sciences. Take the fact that water is a liquid. Well, an individual water molecule isn't liquid, but when you put a lot of them together, suddenly you have water, and that is liquid. Liquidity has *emerged* at a larger scale. But a claim that one thing emerges from another needs careful scrutiny. The emergent liquidity of water can be explained by what we already know about individual water molecules and how they interact. There is no need for some extra, magical ingredient that imports the liquidity. But there does seem to be a missing ingredient for the claim that, from the workings of individual neurons, first-person experience emerges. We can't begin to explain how that might work. So Strawson names this kind of emergence *radical emergence*. 'Spooky!' he calls it.

The radical-emergence view is a black-box idea. Yes, there might be a mechanism in this black box that we don't currently know about. But it would have to be a very special one – an unprecedented one, in fact. A mechanism like none we've ever encountered before, by virtue of which purely physical events can produce the mental experience you are having at this moment. No one has ever had any decent idea – even in principle – about what such a mechanism might be like.

'There's no further argument,' Strawson admits here. If you believe that the mind radically emerges from the neurons, so be it. 'I can't argue further forward for a rejection of radical emergence. It's very much a choice point.' You could decide to say that, since we know we are conscious, then radical emergence *must* happen, even though we don't know how. But if you feel that radical emergence is too spooky, then you are still on the path with Strawson. Now he'll ask a good question. 'Why are you committed to believing in non-experiential stuff?' That is, why do you think that physical matter is *not* conscious?

'And then my favourite line is,' Strawson continues gleefully, '"What's the evidence for that?" Answer: "There isn't any! Nor will there ever be!"'

Forced mate

There's no evidence, Strawson argues, for believing that consciousness is *not* a fundamental property of matter. Why do we assume that it isn't? It's an unexamined part of our common sense, baked in so deep that we rarely even notice we are assuming it. Arguably this idea that matter must be unconscious is a hangover from the new, 'corpuscularian' theory of matter that arose at the birth of modern science in the sixteenth and seventeenth centuries. It was then thought that matter ultimately consisted of tiny solid particles (corpuscles), crashing around into one another like billiard balls on a table. Human beings had immaterial souls, but the rest of reality was a giant clockwork mechanism.

But that picture of matter is long outdated now: we have theories of tremendous predictive power that are built on baffling concepts such as wave–particle duality, quantum tunnelling, vacuum energy, and the like. In light of all this, Strawson asks: 'What ground is there apart from just gut – *bad* gut – for saying that matter is ultimately non-experiential? There isn't any.'

Strawson likes to cite the early-twentieth-century astrophysicist Sir Arthur Eddington, who wrote in 1928: 'Science has nothing to say as to the intrinsic nature of the atom.'[10] Science offers reliable mathematical models for how bits of matter interact, but as to their very 'nature', Eddington says; well, physics leaves that 'undetermined and undeterminable'. And Noam Chomsky, more famous for his theories of linguistics and his political writings, also pointed out: 'The mind–body problem can be posed sensibly only insofar as we have a definite conception of body. If we have no such definite and fixed conception, we cannot ask whether some phenomena fall beyond its range.'[11] In other words, we don't know enough about the matter that makes up our bodies to be sure that mind is totally alien to it.

The advances of modern physics, far from nailing down a definite conception of 'body' or matter, have on the contrary seemed intent on dissolving it utterly. Matter is a wave, it's a particle, it's the vibration of a string, it's a wrinkle in space–time, it's somehow interchangeable with energy . . . basically, what philosophers call the 'ontology' of matter (what it *really is*) is still a kind of conceptual Wild West. Everything is ultimately made up of 'this shimmering pattern of charges', Strawson says. And so, he argues, we have no good reason for supposing that consciousness isn't also part of this shimmering pattern – no reason except gut, and prejudice. Or, you might say, dysfunctional common sense.

'So that's the trajectory,' he concludes. 'It's very simple, it's like a forced mate, as they say in chess. Accept real consciousness, and you want to be a physicalist, then you've got the choice: "radical emergence" or "conscious experience is fundamental". It's as simple as that.'

If you find that you have travelled along Strawson's diagram with him to accepting that conscious experience must be fundamental, then congratulations. You are a panpsychist too.

The long-haired view

Panpsychism is a lot more respectable than it used to be even quite recently. Young philosophers would once be discouraged from pursuing the subject if they wanted to get on in academia, but a new volume of papers on the subject is now due from Oxford University Press. Still, Strawson admits, there are a lot of people who think it's 'bonkers'. 'I really should not grow my hair,' he says with a grin, 'because it's a long-haired view. They can immediately discount the view already because it's just the kind of thing that people with long hair say.'

As it happens, panpsychism is coming back outside philosophy too. The continental guru Bruno Latour is part of a panpsychist turn in anthropology and related disciplines. 'Object Oriented Ontology' seeks to theorise what it is like to be a Lego brick, and questions such as 'What does a forest want?' are not considered nonsensical. David Graeber, the anthropologist and author of *Debt: The First 5,000 Years*, has also endorsed a kind of panpsychist position, arguing that a 'principle of play' might govern the action of electrons.[12] And of course, there is the neo-animistic tidying system of Marie Kondo, which encourages us to consider the feelings of our socks. We seem to be living through a panpsychist revival. Everyone's busy putting the mental back into the fundamental.

Panpsychism is not so hippy a view that one is required to believe that atoms, rocks, and tables think and dream. The glimmers of conscious experience could be very faint in objects less complex than human brains. (In any case, we can't ask rocks – or socks – what they think.) But the idea can alter your worldview more subtly. It means, for one thing, that you can never get away from other minds. 'I know one person who feels that he doesn't *want* there to be so much mind around,' Strawson says, smiling. In other words, his friend doesn't want panpsychism to be true. 'He's a Canadian. He says: "I want to walk out into the wilds of northern Canada and be alone. I don't want any *chatter* around me."'

The mind's black box

For sure, panpsychism as a view still faces stiff challenges. Its opponents consider the most profound to be the 'combination problem'. If all the particles that make up my brain themselves have tiny proto-minds, why does my brain as a whole have a rich and complex, unified and 'big' mind? Why doesn't it just remain a loose coalition of almost-completely-stupid particles? We know from our own lives that lots of little pains don't sum – they don't add up to one big pain. They stay small and separate. So surely lots of little thin consciousnesses can't add up to one rich complex consciousness. Experiences don't sum. That's the combination problem.

Writing in the late nineteenth century, the psychologist William James said that panpsychism was 'almost irresistibly attractive to the intellect', but considered that the combination problem torpedoed it as a serious idea. Yet he couldn't abandon it, and two decades later he decided that because panpsychism *must* be true, combination must somehow be possible.[13]

For his part, Strawson suspects that the combination problem is not a problem, because of what sounds like an even more radical idea. We imagine that there are lots of distinct things in the world – countless subatomic particles like very tiny billiard balls. Strawson dubs this view *smallism*. Well, what if smallism isn't true? What if there just *aren't* a lot of very small things? Modern physics doesn't actually picture an electron as a tiny billiard ball, but as an excitation of a field. A field is a fluid medium that pervades space. And fields *can* sum, mathematically speaking, in a way that perhaps experiences can. There are historical antecedents for the rejection of smallism. The seventeenth-century Dutch lens-grinder and philosopher Baruch Spinoza, who was also a panpsychist, considered that the entire universe was made up of one single substance, which was the same thing as God or Nature. So for him there was no mind–body problem either: what we call mental and physical reality are just the two aspects of this single substance that human beings are able to perceive. And what seem to us to be many different things are also just different modes of this one universe-substance, which has infinitely many attributes.[14] So smallism is an illusion. (It follows, further, that on Spinoza's view there are not many separate minds, either, but just one: the mind of God, in which we all share. The same idea is found in Vedantic and Buddhist philosophy, and was endorsed in the twentieth century by the physicist Erwin Schrödinger.)[15]

So the position of the modern panpsychist is this. If smallism is true, then combination must be possible – because here you and I are, having conscious experiences. And if smallism isn't true, then there's some kind of fusion of fields going on, which enables us, again, to be here having conscious experiences. It's 'impressionistic', Strawson admits, 'but I don't think anyone's much further ahead than that'.

Lest this sound by now the sort of abstract view that only a long-haired philosopher would hold, Strawson points out that he has allies among people who actually poke around in brains for a living. The neurosurgeon Henry Marsh, author of the book *Do No Harm*, has told Strawson that he feels he is probably a panpsychic too. Modern neuroscience, like modern physics, is no barrier to panpsychism, and arguably makes it more credible than ever before.

But just how credible? The way Galen Strawson's argument for panpsychism is structured, we are invited to accept it because other possible views are rejected. But could there ever be some kind of positive evidence in favour of panpsychism?

'I'm not sure what that would be,' Strawson says, musing, and then pounces. 'Well, the first point is: formally speaking, you get the problem of other minds,' he says. 'I can't even know that *you've* got a mind!'

This is true. I could just be a cleverly constructed unconscious robot. Strawson has no access to my conscious experience; he has to infer that I do have a mind by observing my behaviour. So if Strawson can't be sure that I, a person sitting opposite him and speaking apparently relevant sentences in response to his own, actually have a mind, then how much convincing evidence could he ever have that an electron had a mind, even if it did?

There is another possibility for positive support for panpsychism, though. 'If somehow we had a proof that there couldn't be radical emergence,' Strawson says, 'that would do it, I think. Though strictly speaking it would only get you to saying *at least some* particles must already involve experience. It wouldn't formally get you all the way.'

But if you thought that *some* of the fundamental particles involve experience, then you'd have to explain why some do and some don't, right?

'You would,' Strawson agrees. 'And then you get this idea that I call fungibility.' If a type of thing is fungible, then one instance of it will do as well as any other. The classic example is money: you don't need *this particular* ten-pound note to pay for your drinks; any ten-pound note will do. And perhaps matter itself is fungible. 'As far as we know,' Strawson says, 'any form of matter can be converted into any other form. If that's the case, then that means you could build a brain out of any matter.' (You could in principle build a brain out of a piano simply by taking all the constituent electrons and so on and rearranging them.) In which case it looks as if either none of the particles involve experience, or all of them do. And if all of them do, that's panpsychism.

'The conclusion is qualified,' Strawson points out, having finished his diagram, 'because it just says it's the theoretically best-looking position.'

You mean that it seems the least ridiculous option?

'Yes!' he says. 'And deep, deep prejudice is all that stands in the way of seeing that. Just a gut feeling.'

So panpsychism is the last theory standing, as it were, the only one that doesn't commit us to silliness or spookiness?

'Exactly,' Strawson says.

It is a perfectly respectable rule of thumb to accept the least-wrong-looking option. After all, remember what Sherlock Holmes said: 'when you have eliminated the impossible, whatever remains, *however improbable*, must be the truth.' Though you wouldn't suspect this from the ferocity of the attacks on another prominent modern philosopher by his scientific critics.

The ends of the Earth

In 2012, one of the world's most famous philosophers decided to revive another idea that had long been consigned to the dustbin of scientific history. But, unlike panpsychism, this one had had hardly any defenders at all for half a millennium. Our old chicken-freezing friend Francis Bacon decried this idea nearly four hundred years ago. 'Like a virgin consecrated to God,' he wrote, it 'produces nothing.' From then on it was widely considered anti-rational nonsense, the last resort of unfashionable idealists and religious agitators.

And then in 2012 a modern philosopher, Thomas Nagel, published a book called *Mind and Cosmos*, arguing that we should take it seriously after all. Biologists and philosophers were outraged, and lined up to give the malefactor a kicking. His ideas were 'outdated', complained some. One wrote: 'I regret the appearance of this book.'[16] On Twitter, Steven Pinker sneered at 'the shoddy reasoning of a once-great thinker'. It was clearly 'the most despised science book of 2012'.[17] So what had made everyone so angry? It was that someone had dared to take seriously the banned concept of teleology.

In ancient science (or, as it used to be called, natural philosophy), teleology held that things – in particular, living things – had a natural end, or *telos*, at which they aimed. The idea first appeared in Plato, and was developed in relation to the natural world by Aristotle, who argued that the acorn, for example, sprouted and grew into a seedling because its *purpose* was to become a mighty oak. (We would now say that the acorn's DNA contains all the instructions necessary to build an oak tree, but there is no 'picture' of the final tree encoded in the genes.) Sometimes, teleology seemed to imply an intention to pursue such an end, if not in the organism then in the mind of a creator. It also implied reverse causation, with the *telos* – or 'final cause' – acting backwards in time to affect earlier events. For us, the oak tree that the acorn will become cannot be the *cause* of the acorn's growth, for it doesn't exist yet – and when it does exist, it won't be able to cause something that happened beforehand. For such reasons, teleology was ceremonially disowned at the birth of modern experimental science.

The extraordinary success since then of non-teleological scientific thinking and its commitment to forwards-only 'mechanistic causation' would seem to support Francis Bacon's denunciation of teleology. But it continued to bubble under the surface as a live problem for some, particularly regarding descriptions of life. Immanuel Kant wrote that, when observing a living being, we couldn't help thinking in teleological terms – to say, for example, that the *purpose* of the eye is to enable the animal to see – and this kind of thinking was justified for its scientific usefulness. Even so, he concluded, an ultimately teleological explanation was impermissible, since we could never know whether it was true or not. Friedrich Engels hailed the publication in 1859 of Darwin's *Origin of Species* as the final nail in the coffin for teleology; yet one of Darwin's admirers, the American botanist and champion of evolution Asa Gray, felt able to read it nonetheless as confirming a teleological view of life's development, a position that Darwin himself (mostly) rejected.

But the idea that living things had purpose just would not go away. And it sparked developments in fields beyond biology. Norbert Wiener, in his classic 1948 treatise *Cybernetics: Or Control and Communication in the Animal and the Machine*, argued that once artificial systems are engineered to include feedback (when the output becomes part of the next input), then we have created a new kind of 'teleological machine': a machine that has purpose, just like an organism does. Later, the philosopher Alasdair MacIntyre's *After Virtue* (1981) argued that moral philosophy had lost its way precisely because it had abandoned Aristotelian teleology – the idea that there was an essential 'true end' for a human being, a naturally correct way for a person to flourish. 'The whole point of ethics,' MacIntyre wrote, 'is to enable man to pass from his present state to his true end.'[18] If you no longer believe in such a true end, he argued, the whole enterprise lacks rational grounding.

In *Mind and Cosmos*, subtitled 'Why the Materialist Neo-Darwinian Conception of Nature is Almost Certainly False', Thomas Nagel revives the concept of teleology on the basis of his conviction that the mind–body problem has more serious ramifications for evolutionary science than is ordinarily accepted. Nagel says that the appearance of conscious beings such as us can be described as the universe waking up. Yet to him it seems unlikely that life would ever have got started in the first place by somehow springing forth from 'dead matter'; still more unlikely that some forms of life would have developed consciousness; and extremely improbable that one form of life would have acquired the 'transcendent' power of reason.[19] In order to explain these events, Nagel suggests, you need more than simply the 'mechanistic' tools of the laws of physics, natural selection, and so on. You need not just physical theory but 'psychophysical theory'.[20] And you might even need teleology.

It's a bold claim, but not in itself an unscientific one. What really infuriated his critics was Nagel's doubtful attitude towards evolutionary biology. 'It is prima facie highly implausible,' he intuits, 'that life as we know it is the result of a sequence of physical accidents together with the mechanism of natural selection.' There are insuperable 'questions of probability'. The current orthodoxy, he suggests, 'flies in the face of common sense'.[21]

Yet it has always been the job and the glory of science to fly in the face of common sense. If a theory that is robustly supported by observation conflicts with your common sense, you had better adjust the latter. And so we now accept, for example, that apparently solid objects are composed of atoms that are largely made up of empty space – just a shimmering pattern of charges – and that the Earth goes round the Sun and not vice versa. Maybe the appearance of consciousness in the universe was, as Nagel thinks, highly improbable. But improbable is not impossible. Indeed, it sounds odd to say of an event in the past that it was improbable. We know it happened. Therefore its probability *now* of having happened is 100 per cent. So it goes.

Such objections do not rule out teleology altogether. What Nagel's critics rarely conceded was the fact that teleological talk remains rampant to this day in popular and even academic science writing. Vast subterranean seams of purposive metaphor imply a picture of final cause not only in modern biology but in chemistry and physics, too. It has long been accepted that ordinary descriptions of biological function, such as 'The heart is *for* pumping blood', are a kind of shorthand that isn't meant to imply a real commitment to teleology. But we also commonly read, for example, that subatomic particles 'know' or 'choose' the 'right' path to take; that molecules rearrange themselves 'in order to' achieve a certain energy state; or that traits in organisms evolve 'in order to' allow the animal to do something new. Everywhere, indeed, we ascribe purpose to the workings of the natural world.

It's possible to bite the bullet at this point, take all the purpose talk literally, and embrace panpsychism. But Nagel, while he acknowledges panpsychism's attractions – in 1979 he published one of the first respectable modern philosophical papers on the topic – now chooses to cross a different bridge. Teleology, he says, must be at work: a law-like tendency in the universe that loads the dice in favour of the appearance of consciousness. According to this idea, Nagel writes, 'things happen because they are on a path that leads to certain outcomes'. It might be that laws of nature are 'biased toward the marvellous'.[22] If so, it would not be surprising that consciousness – something marvellous – appeared, because we live in a universe whose very purpose, aim, or *telos*, is the production of consciousness.

Other fundamental questions about the distant past might also be amenable to teleological therapy. Indeed, teleological principles are these days being seriously proposed in cosmology by some people as an answer to the 'fine-tuning' puzzle – why the laws of nature are just right, to a very fine degree of precision, for allowing a universe that can support life. The physicist Paul Davies endorses a teleological 'life principle' in his book *The Goldilocks Enigma* (2007). Meanwhile, the philosophers John Hawthorne and Daniel Nolan show how teleological laws in principle could even explain why there would be any universes at all in the first place, or 'why there is something rather than nothing' – a question that is otherwise, depending on one's philosophical and scientific proclivities, either irrelevant, deeply mysterious, or nonsensical.[23]

Fundamental teleology can be respectably revived by a modern thinker as brilliant as Thomas Nagel because it is not an unscientific hypothesis. We don't know how to test for it, however, or have much idea of what teleological laws might say. Nagel himself doesn't attempt a detailed description of teleological laws, leaving the task hopefully to creative scientists of the future. If it's true, then teleology is for the moment another black box.

Eternal recurrence

It looks as though ideas recur in philosophy more frequently than in other disciplines. Perhaps this is because, although there is certainly progress in the understanding of certain philosophical problems, there is rarely a knock-down argument that everyone accepts for the truth of one position rather than another. (By contrast, mathematics proceeds exactly by way of knock-down arguments, or proofs: Euclidean geometry, for example, still works for flat planes, whereas the idea that the planets move in perfect circles has been definitively refuted.) Many philosophers talk of finding that what they supposed to be their original thoughts had been anticipated by someone else, often long in the past. Some people might find this annoying, but in a lovely passage in one of his papers Galen Strawson writes that he has the opposite feeling:

> My experience since I first lectured on the 'mind–body problem' in the late 1980s has been one of finding, piece by piece, through half-haphazard reading, that almost everything worthwhile that I have thought of has been thought of before, in some manner, by great philosophers in previous centuries (I am sure further reading would remove the 'almost'). It is very moving to discover agreement across the centuries, and I quote these philosophers freely, and take their agreement to be a powerful source of support. Almost everything worthwhile in philosophy has been thought of before, but this isn't in any way a depressing fact [. . .] and the local originality that consists in having an idea oneself and later finding that it has already been had by someone else is extremely common in philosophy, and crucial to philosophical understanding.[24]

When I remind him of this passage, Strawson says: 'Oh yes. I enjoy that.' In general, he says: 'I just want to assert that there are certain human universals. Not just across cultures synchronically, but also *through* time: the same problems keep coming. And it's not just philosophy either. When I read Aristophanes, he's *really funny* – it is moving, and it makes me laugh.'

Pointing out that philosophical ideas have been had before can become a kind of sport. 'I was a member of Freddy Ayer's discussion group for a long time in Oxford,' Strawson remembers, 'and it soon became clear to me that there was a distinctive role for the senior members, which was basically [to say] "I've heard this before. It's new, with new knobs, but . . ." That certainly happened to me too. But – okay, it's come round before, but maybe it's important that people should sort of feel it's new and exciting.'

And if it's come round before then maybe there's something in it after all?

'Well, indeed! That's right, if it's an actual claim, but of course often it's *thinking there's a problem* that's come round before – oh, I've got another quotation!' He is peering intently into his MacBook Air. 'I'm just a quotation machine,' he says happily. 'This is one of my favourites.'

It is Schopenhauer: 'To truth, only a brief celebration of victory is allowed between the two long periods in which it is condemned as paradoxical or disparaged as trivial.'[25]

'Things,' Strawson says, 'we get them right, and then we lose them again. That's kind of tragic.'

But then we can find them again. All the more reason to feel a frisson upon discovering allies in the past who got things right the first time, to be 'moved by the sense of connection across the centuries'. Rethink, it turns out, can be an emotional business.

The view from the bridge

When ideas are resurrected, it is not necessarily because everyone can now see that they are true. Teleology might inform the laws of nature or it might not, but Nagel, at least, argues that it supplies something that is missing from present-day explanations. For him, it goes some way towards bridging the explanatory gap about how consciousness arises at all. And for teleologically-minded cosmologists, the idea helps to bridge the explanatory gap between the vast possibilities of existence and the fact that we happen to find ourselves in just this particular universe. Galen Strawson, meanwhile, saw that a different old theory simply dissolved the mind–body problem, and that the best modern science arguably made it look more attractive than it had done centuries ago.

Both thinkers knew what they were doing: consciously turning over an old idea and seeing how it shone in the light of present-day understandings. They were rethinking old ideas in order to bridge a gap in the modern worldview. And in doing so they exploit the power of rethink to challenge our common sense, and to encourage us to question a gut feeling that might just be bad. Their solutions seem unlikely, but the alternatives may seem unlikelier still.

To spend such energy addressing questions that we may never be able to answer definitively might seem to some more practical-minded people a regrettable diversion of intellectual force. On this question I like the pragmatic defence of such inquiry offered by Thomas Malthus, who became notorious in the nineteenth century for his views on the increase in human population – but much more about that later. 'It is probable,' Malthus wrote, 'that man, while on earth, will never be able to attain complete satisfaction on these subjects, but this is by no means a reason that he should not engage in them. The darkness that surrounds these interesting topics of human curiosity may be intended to furnish endless motives to intellectual activity and exertion.' Imagine, Malthus suggests, that a 'Supreme Being' suddenly gave us, in irrefutable and totally credible fashion, an explanation of 'the nature and structure of mind [. . .] and the whole plan and scheme of the Universe'. Such a revelation, he argues, would give people nothing left to strive for in the intellectual realm, and so it 'would act like the touch of a torpedo on all intellectual exertion and would almost put an end to the existence of virtue'.[26]

The lure of the darkness is precisely what keeps us thinking at all.

Is there something about the secular modern West's restless search for authenticity and meaning that helps explain why these ideas are coming back now? In other areas of culture the new is increasingly dressed up as old: jeans with ready-made holes, artificially distressed and 'road-worn' electric guitars, or modern reconstructions of supposedly ancient religions, such as Wicca.[27] Perhaps the rehabilitation of notions like panpsychism and teleology is the intellectual equivalent of the desire, among some environmentalists, for 'rewilding' – the attempt to recreate earlier ecosystems by reintroducing wolves and other large animals to the countryside – or even for 'de-extinction', the dream of cloning woolly mammoths and the like from preserved fragments of DNA, the fictional warning of *Jurassic Park* notwithstanding. When we pivot from unsatisfying alternatives, do we seek the imprimatur of history after all? Perhaps such philosophical rethinking is a kind of neoclassicism.

And yet it is also motivated by the latest discoveries of science, such as the puzzlingly immaterial nature of matter. In this way, new ideas can make ancient intuitions even more attractive than they were before. It is not as though teleology and panpsychism are now concretely vindicated, like the idea of an electric car or a programming language. Nor, however, are they just irrelevant metaphysical speculation: things out there in the world must, after all, be one way or the other. Ideas such as these keep returning throughout history to scratch a reality itch that we cannot otherwise reach. They nag at, and help to define, the limits of our understanding. In the mean time, you might begin to change the way you put away your socks.

8
When Zombies Attack

And sometimes, ideas return when they definitely should have stayed dead.

'If fifty million people say something foolish, it is still foolish.'
– Somerset Maugham

Rotters

So far we have considered the revival of many ideas that have been brilliantly upgraded or repurposed, or seem newly compelling to the modern age. Some ideas from the past, on the other hand, are just dead wrong: surely *they'll* never come back. And yet they do. That can be funny; but it's also a problem.

In January 2016, the rapper B.o.B took to Twitter to drop science on his fans about how the Earth is really flat. 'A lot of people are turned off by the phrase "flat earth"', he acknowledged, 'but there's no way u can see all the evidence and not know . . . grow up.' At length the astrophysicist Neil deGrasse Tyson joined in the conversation, offering friendly corrections to B.o.B's zany proofs of non-globism, and finishing with a sarcastic compliment: 'Being five centuries regressed in your reasoning doesn't mean we all can't still like your music.'

Actually, it's a lot more than five centuries regressed. Contrary to what we often hear, people didn't think the Earth was flat right up until Columbus sailed to the Americas. It was already obvious to all educated people that it was round. (Although not exactly a globe: in the fifteenth century the model was that the land and oceans formed a bulging agglomeration of two spheres.)[1] Magellan had circumnavigated the world in 1519–22, but people had already long been convinced of its basic shape. In ancient Greece, the philosophers Pythagoras and Parmenides had already recognised that the Earth was spherical. Plato said it was a round body in the middle of the heavens. Aristotle pointed out that you could see some stars in Egypt and Cyprus that were not visible at more northerly latitudes, and also that the Earth casts a curved shadow on the Moon during a lunar eclipse. The Earth, he concluded with impeccable logic, must be round. From then on, the only people who thought the Earth was flat were certain Christian theologians who stuck to biblical literalism, but even this current expired in the eighth century CE. Ever since, the flat-earth view has been simply ridiculous. Until very recently, with the resurgence of apparently serious Flat Earthism on the internet. And it's not just one rapper. An American named Mark Sargent, formerly a professional videogamer and software consultant, has had millions of views on YouTube for his 'Flat Earth Clues' video series. ('You are living inside a giant enclosed system,' his website warns.)[2] Another prominent Flat Earther is a former NASA artist and stand-up comic, Math Boylan. And the Flat Earth Society is alive and well, with a thriving website.

This is highly peculiar because we all know the Earth isn't flat, and indeed we reserve the term 'Flat Earthers' for people who believe utter nonsense. In December 2015, the physicist and writer Brian Cox tweeted: 'Most surprising thing I discovered this year is that there are actually people alive today who think the Earth is flat. Genuinely baffled!' So what, if anything, explains it?

The dark side of the fruitful reconsideration of old ideas we have seen in this book is that some old ideas that are revived really should have been left to rot. What is rediscovered is a shambling corpse. These are *zombie ideas*. You can try to kill them, but they just won't die. We're going to meet more of them soon. And their existence is a big problem for our normal assumptions about how the marketplace of ideas operates.

What exactly is the 'marketplace of ideas'? The phrase was originally used as a way of defending free speech. Just as traders and customers are free to buy and sell wares in the market, so freedom of speech ensures that people are free to exchange ideas, test them out, and see which ones rise to the top. So far, so good. But there's more to it than that. The modern notion of the marketplace of ideas assumes that the ideas that rise to the top will be the *best* ones. We are encouraged to believe that all contests between ideas will be resolved to everyone's benefit by competition in the marketplace. Just as good consumer products succeed and bad ones fail, so in the marketplace of ideas the truth will win out, and error and dishonesty will disappear.

There is certainly some truth in the thought that competition between ideas is necessary for the advancement of our understanding. But the economic metaphor of the marketplace obscures some important counter-intuitive ways in which this actually happens, and is unable to account for the surprising prevalence of market failures. The idea that the best ideas will always succeed is rather like the idea that unregulated financial markets will always produce the best economic outcomes. In each case the assumption is that truth, or perfect economic efficiency, will emerge automatically from the operations of the market – *some day*. Just because it hasn't happened yet, so its adherents plead, doesn't mean it won't happen at some undefinable point in the future. As IMF chief Christine Lagarde put this standard wisdom laconically in Davos: 'The market sorts things out, eventually.'[3] Maybe so. But while we wait, very bad things might happen. And one of those bad things is the recurrence of zombie ideas.

Zombies don't occur in physical marketplaces – say, the market for technology products. No one now buys Betamax video recorders, because that technology has been superseded and has no chance of coming back. (The reason that other old technologies, such as the manual typewriter or the acoustic piano, are still in use is that according to the preferences of their users they haven't been superseded.) So zombies such as Flat Earthism simply shouldn't be possible in a well-functioning marketplace of ideas. And yet – *they live*. How come?

One clue is provided by economics. It turns out that the marketplace of economic ideas is itself densely infested with zombies. After the financial crisis had struck, the Australian economist John Quiggin published an illuminating work called *Zombie Economics*, describing economic theories that still somehow shambled around even though they were clearly dead, having been refuted by actual events in the world. An example is the notorious 'Efficient Markets Hypothesis', which holds (in its strongest form) that 'financial markets are the best possible guide to the value of economic assets and therefore to decisions about investment and production'.[4] That, Quiggin argues, simply 'can't be right' – since over the past decade and a half, 'global financial markets have shown themselves subject to the same manias, bubbles, and busts that were seen in the Dutch tulip craze of the seventeenth century'.[5] Not only was the Efficient Markets Hypothesis refuted by the global meltdown of 2007–8, in Quiggin's view it actually caused it in the first place: the idea 'justified, and indeed demanded, financial deregulation, the removal of controls on international capital flows, and a massive expansion of the financial sector. These developments ultimately produced the Global Financial Crisis'.[6]

What can explain the persistence of such zombies? Well, one answer is that an idea will have a good chance of hanging around as a zombie if it benefits some influential group of people. The Efficient Markets Hypothesis is financially beneficial for bankers who want to make deals unencumbered by regulation. A similar point can be made about the privatisation of state-owned industry: it is hardly ever good for citizens, but is always a cash bonanza for those directly involved. The same goes, Quiggin argues, for the idea of 'trickle-down economics', the view that making the best-off people even better off will be good for everyone. 'Clearly,' Quiggin observes drily, 'an idea so appealing to people who can afford to reward its promulgators is unlikely to be killed by mere evidence of its falsehood.'[7]

Yet the metabolism and reproductive life of your average zombie is still little understood. Other zombie ideas do not seem directly to enhance the status of the powerful, so we must look for other reasons why they persist. An alternative recipe to make a zombie is to take one part snowball effect (as a piquant-looking idea gets repeated) and mix with one part laziness (as no one thinks to check the original source).

You probably know that the human tongue has 'regional sensitivities' – sweetness is sensed on the tip, saltiness and sourness on the sides, and bitter at the back. At some point you've seen a scientific 'tongue map' showing this – they appear in cookery books as well as medical textbooks. It's one of those nice, slightly surprising findings of science that no one questions. And it's rubbish.

As the eminent professor of biology, Stuart Firestein, explains, the tongue-map myth arose because of a mistranslation of a 1901 German physiology textbook. Regions of the tongue are just 'very slightly' more or less sensitive to each of the four basic tastes, but they each can sense all of them. The translation 'considerably overstated' the original author's claims. And so 'the mythical tongue map was canonized into a fact and maintained by repetition, rather than experimentation, now having endured for more than a century'.[8]

Just as it would be alarming to meet a friend for a drink and find that she had turned into a zombie, it is alarming to discover that even some widely accepted things are secret zombies, having somehow fooled everyone for years by wearing normal clothes and not biting people. 'The more successful the fact,' Firestein warns, 'the more worrisome it may be. Really successful facts have a tendency to become impregnable to revision.'[9] So: question everything. If the marketplace of ideas can confer authority through mere repetition, as in this example, then experiment is actually inimical to it.

There are some more terrifying zombies than the Efficient Markets Hypothesis or the tongue map. Our world is full of *moral zombies*, too: ideas that once were unthinkable and that came back – but shouldn't have. Take the practice of torture. The historian of technology David Edgerton reports: 'In a history of torture published in 1940 there is not much on the twentieth century, and little if anything that is seen as new.

Of course the Nazis were associated with torture, but they were generally seen as a throwback. Yet the years after the Second World War, far from seeing the retreat of torture, saw its extension and its technological refinement.'[10] Then, after 2001, the Bush–Cheney administration argued openly for the use of torture, under various weasel euphemisms such as 'enhanced interrogation'. What was once barbaric was openly thinkable again. On the other side, and quite possibly as a direct reaction, the practice of killing people by sawing their heads off with a hand-held knife was revived by 'Islamic State' fighters, and their PR effort amply rewarded by breathless media reports and the wide circulation of 'beheading' videos.

Meanwhile, neo-fascist movements are on the rise in Europe. Piracy and slavery are rampant, and thanks to the magic of globalisation can morally implicate societies where they have supposedly been eliminated. In Britain, it was announced in 2015, high-street clothing stores with a turnover of more than £36 million will have to publish annual reports showing what they have done to ensure that slavery has not contributed anywhere in their supply chain.[11]

One of the paradoxes of zombie ideas, however, is that they can have positive social effects. The answer is not necessarily to suppress them, since even apparently vicious and disingenuous ideas can have a productive place in the ecology of human inquiry. As David Edgerton points out, even 'Holocaust denial, more accurately gas chamber denial, has led to research that shows in surprising detail how the SS built and used the gas chambers, weakening even further the denier case'.[12] Few would argue that a commercial marketplace *needs* fraud and faulty products. But in the marketplace of ideas, zombies can actually be useful. Or if not, they can at least make us feel better. That, paradoxically, is what I think the Flat Earthers of today are really offering – comfort.

The world is flat

Today's rejuvenated Flat Earth philosophy, as promoted by rappers and YouTube videos, is not simply a recrudescence of pre-scientific ignorance.

It is, rather, the mother of all conspiracy theories. The point is that everyone who claims the Earth is round is trying to fool you, and keep you in the dark. In that sense, it is a very modern version of an old idea.

As with any conspiracy theory, the Flat Earth idea is introduced by way of a handful of seeming anomalies, things that don't seem to fit the 'official' story. Have you ever wondered, the Flat Earther will ask, why commercial aeroplanes don't fly over Antarctica? It would, after all, be the most direct route from South Africa to New Zealand, or from Sydney to Buenos Aires – *if* the Earth were round. But it isn't. There is no such thing as the South Pole, so flying over Antarctica wouldn't make any sense. Plus, the Antarctic Treaty, signed by the world's most powerful countries, bans any flights over it, because something very weird is going on there. So begins the conspiracy sell. Well, in fact, some commercial routes do fly over part of the continent of Antarctica. The reason none overfly the South Pole itself is because of aviation rules that require any aircraft taking such a route to have expensive survival equipment for all passengers on board – which would obviously be prohibitive for a passenger jet. Most of Antarctica is out of range of the regulations that specify how far away an aircraft may fly from any available emergency-landing airport option. (For comparison, it is routine for aircraft on other long-haul routes to fly almost directly over the North Pole, which is considerably less remote from civilisation.)[13]

Okay, the Flat Earther will say, then what about the fact that photographs taken from mountains or hot-air balloons don't show any curvature of the horizon? It is perfectly flat – therefore the Earth must be flat. Well, a reasonable person will respond, it *looks* flat because the Earth, though round, is really very big. But in fact the visible horizon *is* faintly curved. Photographs taken from the International Space Station in orbit show a very obviously curved Earth.

And here is where the conspiracy really gets going. To a Flat Earther, any photograph from the ISS is just a fake. So too are the famous photographs of the whole round Earth hanging in space that were taken on the Apollo missions. Of course, the Moon landings were faked too. This is a conspiracy theory that swallows other conspiracy theories whole. (I haven't found a Flat Earther claiming that JFK was assassinated because he was about to reveal to the public that the world wasn't round, but I wouldn't be surprised.) According to Mark Sargent's 'enclosed world' version of the Flat Earth theory, indeed, space travel had to be faked because there is actually an impermeable solid dome enclosing our flat planet. The US and USSR tried to break through this dome by nuking it in the 1950s: that's what all those nuclear tests were really about.

Commercial aviation is not faked, according to Flat Earthers (it is hard to see how that would be possible), but GPS systems are secretly rigged to fool pilots into thinking they are flying above a round planet. It turns out, indeed, that in order to maintain belief in a Flat Earth you have to jettison not only trust in electronic instruments but an awful lot of established science. Astronomy will have to go, so some Flat Earthers decide that the Moon and the stars aren't real anyway: they are holographic projections. Gravity (which would pull the Earth into a spherical shape) doesn't exist. (What we feel as gravity is the whole Earth accelerating 'upwards' in space.) And what about sunsets? According to Flat Earth theory the sun is essentially a giant spotlight shining downwards on to a disc. What is obscuring more and more of the Sun while it is setting, if it is not the curvature of the Earth? The Flat Earther just says this is a difficult scientific mystery and requires more investigation.

The intellectual dynamic here, then, is one of rejection and obfuscation. The Flat Earther can easily dismiss things like photographs of the Earth from space as fakes and simply refuse to follow elementary geometrical demonstrations, using shadows cast by sticks in different places, that were known in antiquity to prove the Earth's roundness. A better way to challenge such views might relate to their physical and psychological consistency. If the Earth is flat, has anyone ever fallen off? Why don't we know of any cities right on the edge? What keeps the oceans from spilling over the sides? But there are answers to these questions too. Remember the Antarctic? Well, it's not a continent at all. It's a massive, impenetrable ice wall around the perimeter of the Earth. That would be why you can't fly over it.

A lot of ingenuity evidently goes into the elaboration of modern Flat Earth theories to keep them consistent. It is tempting to suppose that some of the leading writers (or, as fans call them, 'researchers') on the topic are cynically having some intellectual fun, but there are also a lot of true believers on the messageboards who find the notion of the 'globist' conspiracy somehow comforting and consonant with their dark idea of how the world works. (There is even a metaconspiratorial subcurrent of people who believe that the leading Flat Earth proponents, such as Mark Sargent, are disinformation agents trying to make the truth seem ridiculous.) You might think that the really obvious question here, though, is: what purpose would such an incredibly elaborate and expensive conspiracy serve? Who benefits if the world's population are deceived into thinking the Earth is round? What exactly is the point?

It seems to me that the desire to believe such stuff stems from a deranged kind of optimism about the capabilities of human beings. It is a dark view of human nature, to be sure, but it is also rather awe-inspiring to think of secret agencies so single-minded and powerful that they really can fool the world's population over something so enormous. Certainly another motivation for Flat Earthers is simply aesthetic. Like Hannibal in *The A-Team*, they just love the way a (made-up) plan comes together. 'I was, and still am, a huge conspiracy guy,' confesses Mark Sargent on his website. 'I literally ran out of new tin hat topics to research and I STILL wouldn't look at this one without embarrassment, but every time I glanced at it there was something unresolved, and once I saw the near perfection of the whole plan, I was hooked.' It *is* rather beautiful. Bonkers, but beautiful. As the much more noxious example of Scientology also demonstrates, it is all too tempting to take science fiction for truth – because narratives always make more sense than reality.

Conspiracy markets

We know that it's a good habit to question received wisdom, and to be alert to the possibility of something like a 'tongue map' being a myth. Sometimes, though, healthy scepticism can run over into paranoid cynicism, and giant conspiracies seem oddly consoling. Suddenly, the world can look flat.

When Nasa's New Horizons mission sent back the first close-up photographs of Pluto in July 2015, a cadre of 'Pluto truthers' attempted to demonstrate that the images were faked. It's interesting that such people call themselves 'truthers' – after the original '9/11 Truth' movement – when in fact they are in the business of declaring that things are *untrue*, or fake. We should probably call them Falsers.

The 9/11 Falsers – along with the modern Flat Earthers – are a good example of another kind of market failure in the marketplace of ideas: that wrong ideas can go viral just as quickly, if not more so, than accurate ones – thanks to what social scientists call 'informational cascades' and echo-chamber effects. (And thanks to the simple frisson of a good story. Hence urban myths such as the person who wakes up in an ice-cold bath to see a note informing them that their kidneys have been removed. As far as anyone can tell, such a 'Kidney Heist' by unscrupulous organ-traders has never actually happened.)[14] As Cass Sunstein observes: 'The marketplace of ideas will often fail to produce the truth' because social cascades and group polarisation actually '*ensure* that any marketplace will lead many people to accept damaging and destructive falsehoods' (my emphasis).[15] Added to those factors is what we might call the lamentable persistence of media in the information age: the YouTube 9/11-conspiracy film, *Loose Change*, still attracts and impresses new viewers, even though all of its claims have been refuted in exhausting detail elsewhere.

One reason why myths and urban legends hang around so long seems to be that we like simple explanations and are inclined to believe them. The 'MMR causes autism' scare perpetrated by Andrew Wakefield, for example, had the apparent virtue of naming a concrete cause (vaccination) for a deeply worrying and little-understood syndrome (autism). Years after it was shown that there was nothing to Wakefield's claims, there is still a strong and growing 'anti-vaxxer' movement, particularly in the US, which poses a serious danger to public health.

The benefits of immunisation, it seems, have been forgotten and might need to be relearned, at enormous cost in suffering. In the autumn of 2015, public-health officials in Ukraine feared a major outbreak of polio, because public distrust in immunisation had seen the take-up of polio jabs fall to only 14 per cent of children.[16]

The yearning for simple explanations also helps account for the popularity of outlandish conspiracy theories that paint a reassuring picture of all the world's evils as being attributable to a cabal of super-villains. Perhaps a secret society really is running the show – in which case the world at least has a weird kind of coherence. (For many indefatigable scholars, lyrical hints and occult symbols in music videos prove that Jay Z, Beyoncé, Kanye West, and Lady Gaga are all Satan-worshipping members of the Illuminati.)[17] Or take the global anti-Semitic conspiracy theory laid out in the forged *Protocols of the Elders of Zion* that began to be circulated in the late nineteenth century – and still circulates as a zombie in much of the Middle East today – as well as Hitler's subsequent explanation of all Germany's woes as attributable to the machinations of Jews. Of course, these ideas were simply false, but a higher-level lesson is that they explained too much, too easily.

Another widespread conspiracy that operates in our time, however, shows balefully what happens when the world of ideas really does operate as a marketplace. It happens to be the case that many prominent 'climate sceptics' have been secretly funded by oil companies.[18] The idea that there is some scientific controversy over whether burning fossil fuels has contributed in large part to the present global warming (there isn't) is an idea that has been literally bought and sold, and remains extraordinarily successful. That, of course, is just a particularly dramatic example of the way all Western democracies have been captured by industry lobbying and party donations, in which friendly consideration of ideas that increase the profits of business is simply purchased, like any other commodity.

If the marketplace of ideas worked as advertised, not only would this kind of corruption be absent, but it would be impossible in general for ideas to stay rejected by the marketplace for hundreds or thousands of years before eventually being vindicated. Yet, as we have seen over the course of this book, that has repeatedly happened.

The truth is out there

While the return of Flat Earth theories is silly and rather alarming, it does illustrate quite vividly some real and deep issues about human knowledge. How, after all, do you or I know that the Earth really is round? Essentially, we take it on trust. We may have experienced some common indications of it ourselves (seeing only the upper part of the mast that belongs to a far-away ship, because the curved Earth is obscuring the rest of it) – but we accept others' explanations. 'For 20 generations people believed that the earth was round because there was a globe in every classroom they sat in,' writes Mark Sargent. 'There was no proof.' Well, there is and was plenty of proof, but it's true that we don't examine it ourselves: the experts all say the Earth is round; we believe them, and get on with our lives.

The second issue is that we cannot actually know for sure that the way the world appears to us is *not* actually the result of some giant conspiracy or deception. The modern Flat Earth theory comes quite close to an even more all-encompassing species of conspiracy theory familiar from philosophy and theology. It *might* be the case that God created the whole universe, including fossils, ourselves and all our (false) memories, only five minutes ago. Or it might be the case that all my sensory impressions are being fed to my brain by a malign and clever demon intent on deceiving me (Descartes) or by a virtual-reality program controlled by evil sentient artificial intelligences (*The Matrix*).

Even so, it is possible to remain untroubled by this zombie, as by others. We may observe, for instance, that – just as gas-chamber denial spurred further positive research refuting it – the resurgence of Flat Earth theory has also spawned a lot of helpful popular web pages employing mathematics, science, and everyday experience to explain why the world actually is round. This is a boon for public education.

And we certainly should not give in to the temptation to conclude, as some have done, that belief in a conspiracy is prima facie evidence of stupidity. Evidently, conspiracies really happen. Members of al-Qaeda really did conspire in secret to fly planes into the World Trade Center (*pace* the 9/11 Falsers, whose alternative conspiracy theories usually involve Mossad and 'controlled demolition' of the towers). And, as Edward Snowden revealed, the American and British intelligence services really did conspire in secret to intercept the electronic communications of millions of ordinary citizens. Perhaps the most colourful official conspiracy that we now know of happened in China. When the half-millennium-old Tiananmen Gate was found to be falling down in the 1960s, it was secretly replaced, bit by bit, with an exact replica, in a successful conspiracy that involved nearly three thousand people who managed to keep it a secret for years.[19] The idea that conspiracies *don't* happen is a useful sedative for the population and a firewall for the powerful.

Indeed, a healthy conspiriological attitude may even be said to underlie much honest intellectual inquiry. This is how the physicist Frank Wilczek puts it: 'When I was growing up, I loved the idea that great powers and secret meanings lurk behind the appearance of things.'[20] Newton's grand idea of an invisible force (gravity) running the universe was definitely a cosmological conspiracy theory in this sense. Yes, many conspiracy theories are zombies – but so is the idea that thinking in terms of conspiracies is always wrong.

Replicate or perish

Things are better, one assumes, in the rarefied marketplace of scientific ideas. There, the revered scientific journals have rigorous editorial standards and peer review, which help make them a complex but smooth-running institution of knowledge-production. Zombies and other market failures are thereby prevented. Well, not so fast. Remember the tongue map. It turns out that the marketplace of scientific ideas is hardly perfect either.

The psychologist Hermann von Helmholtz saw a serious failure in the marketplace of scientific ideas at the end of the nineteenth century. In describing it, he offers a beautiful thought worthy of the Argentinian fabulist Jorge Luis Borges: 'In the "type case" of the printer,' Helmholtz writes, 'all the wisdom of the world is contained which has been or can be discovered; it is only requisite to know how the letters are to be arranged.' So true – and so little help to us. The problem is that, like Borges's infinite library, the printer's type case also contains limitless amounts of apparently plausible nonsense. Much like, Helmholtz laments, the scientific literature itself. '[I]n the hundreds of books and pamphlets which are every year published about ether, the structure of atoms, the theory of perception,' he says, 'all the most refined shades of possible hypotheses are exhausted, and among these there must necessarily be many fragments of the correct theory. But who knows how to find them?' Who indeed? Helmholtz came to a baleful conclusion: '[A]ll this literature, of untried and unconfirmed hypotheses, has no value in the progress of science. On the contrary, the few sound ideas which they may contain are concealed by the rubbish of the rest.'[21]

It was for such reasons that the scientific community began to adopt the modern institution of 'peer review', in which an article submitted to a journal will be sent out by the editor to several anonymous 'referees' who are expert in the field and will give a considered view on whether the paper is worthy of publication, or will be worthy if revised. (In Britain, the Royal Society had begun to seek such reports in 1832.)[22] The barriers to entry for the best journals in the sciences and humanities mean that – at least in theory – it is impossible to publish 'untried' hypotheses. In our day, peer review is widely considered the gold standard of intellectual seriousness.

But there are increasing rumblings in the academic world itself to the effect that peer review is fundamentally broken. Even that it actively suppresses good new ideas while letting through a multitude of very bad ones. 'False positives and exaggerated results in peer-reviewed scientific studies have reached epidemic proportions in recent years,' *Scientific American* reported in 2011.[23] Indeed the writer of that column, a professor of medicine named John Ioannidis, had previously published a famous paper called 'Why Most Published Research Findings Are False'. The issues, he noted, are particularly severe in healthcare research, in which conflicts of interest arise because studies are funded by large drug companies, but there is also a widely discussed problem in psychology.

Take the widely popularised idea of 'priming'. In 1996, a paper was published claiming that experimental subjects who had been verbally 'primed' to think of old age by being made to think about words such as 'bingo', 'Florida', 'grey', and 'wrinkles' subsequently walked more slowly when they left the laboratory than those who had not been primed. It was a dazzling idea, and led to a flurry of other findings that priming could affect how well you did on a quiz, or how polite you were to a stranger. In recent years, however, researchers have become suspicious, and have not been able to generate the same findings as many of the early priming studies when doing new experiments. This is not definitive proof of falsity, but it does show that publication in a peer-reviewed publication is no absolute guarantee of reliability. We expect robust discoveries to be *replicable* at other times and in other places – but psychology, some argue, is currently going through a crisis in replicability, which Daniel Kahneman has called a looming 'train wreck' for the field as a whole.[24]

Could priming be a future zombie? Well, most people think it's unlikely that all such priming effects will be refuted, since there is now such a wide variety of studies on them. The more interesting problem (and the obvious direction for future research) is to work out what scientists call the idea's 'ecological validity' – that is, how well do the effects translate from the artificial simplicity of the lab situation to the ungovernable messiness of real life? As the philosopher Gary Gutting notes: '[P]riming experiments seldom tell us how important priming is in realistic situations. We know that it has striking effects under highly simplified and controlled laboratory conditions, where the subjects are exposed only to the stimuli that the experimenters provide. But it is very difficult to know how significant priming stimuli (thinking about money, large numbers, abstract questions) would be in a real-life, uncontrolled environment, where all sorts of stimuli might be conflicting with one another.'[25]

All this controversy in psychology just shows science working as it should – being 'self-correcting'. One marketplace-of-ideas problem here, though, is that papers with surprising and socially intriguing results will be described throughout the world's media, and lauded as definitive evidence in popularising books, immediately they are published, and long before awkward second questions begin to be asked. At any given moment, it is really hard to know for sure which studies are reliable and which aren't. In 2012, scientists at the biotech company Amgen, for instance, were able to reproduce only six out of fifty-three 'landmark' cancer studies.[26] The Retraction Watch website, meanwhile, has since 2010 noted hundreds of retractions of scientific articles in journals, for reasons of error and – worryingly common – researcher or reviewer fraud. In other words, the problems with scientific publication that annoyed Hermann von Helmholtz at the end of the nineteenth century are still problems at the beginning of the twenty-first.

It would be sensible, for a start, for everyone to make the apparently trivial rhetorical adjustment from the popular phrase 'Studies show . . .' to something more reflective of the provisionality of hypotheses (and more in line with scientists' own linguistic practice). We could make the effort to limit ourselves to phrases such as 'Studies suggest' or 'Studies indicate'. After all, 'showing' strongly implies *proving*, which is all too rare an activity outside mathematics. Studies are always susceptible to rethinking. That is an indispensable part of their power.

Peer pressure

Nearly every academic inquirer I talked to while researching this book says that the interface of research with publishing is seriously flawed. Partly, for a start, because the incentives are all wrong – a 'publish or perish' culture encourages quantity of published research over quality.

And partly because of the issue of publication bias. This is the name for the well-attested fact that the studies that are published are the ones that yielded the hoped-for results. Studies that failed to show what they hoped languish in desk drawers.

One reform suggested by many people to counteract publication bias would be to encourage the publication of more 'negative findings' – papers where a hypothesis was not backed up by the experiment performed. One problem, of course, is that such findings are not very exciting. Negative results do not make headlines. Part of the problem, actually, is that they are called 'negative results' in the first place. As John Ioannidis remarks, '"Negative" is actually a misnomer.'[27] To find out that something is *not* true can equally well be interpreted as a positive discovery that something we *thought* was true *isn't* – which opens up the field for further investigation. Hence the maintenance by the open-access journal *PLoS* of a list of 'negative' papers. Here we find several failures to replicate priming studies (or, as we might say, successes in finding that priming studies cannot be replicated), as well as a variety of other useful discoveries such as that computer games do not alleviate the symptoms of tinnitus, and that 'Untrained Chimpanzees Fail to Imitate Novel Actions'. This is all good news. 'The publication of negative, null and inconclusive results,' *PLoS* explains, 'is important to provide scientists with balanced information and avoid the duplication of efforts testing similar hypotheses, which waste valuable time and research resources in the process.'

The publication-bias issue is even more pressing in the field of medicine, where it is estimated that the results of around half of all trials conducted – clinical trials, pharmaceutical industry trials, and academic trials – are never published at all, because their results were 'negative'. But this means, as the medical researcher and writer Ben Goldacre explains, that 'we see a biased half of the literature'. 'When half the evidence is withheld,' he writes, 'doctors and patients cannot make informed decisions about which treatment is best.'[28] So Goldacre has heroically kick-started a campaigning group called AllTrials to demand that *all* clinical trials conducted anywhere be registered, with 'their full methods and summary results reported'. That is good news for the health of everyone on the planet.

When lives are not directly at stake, however, it might be difficult to publish more 'negative findings' in other areas of science. One idea, floated by *The Economist*, is that 'Journals should allocate space for "uninteresting" work, and grant-givers should set aside money to pay for it.'[29] It sounds splendid, to have a section in journals for tedious results, or maybe an entire journal dedicated to boring and perfectly unsurprising research. But good luck getting anyone to fund it.

Pushback

The good news is that some of the flaws in the marketplace of scientific ideas might be hidden strengths. It's true that some people think peer review, at its glacial pace and with its bias towards the existing consensus, works to actively repress new ideas that are challenging to received opinion. Notoriously, for example, the paper which first announced the invention of graphene – a way of arranging carbon in a sheet only a single atom thick – was rejected by *Nature* in 2004 on the grounds that doing so was simply 'impossible'. But that idea was too impressive to be suppressed; in fact, the authors of the graphene paper had it published in *Science* only six months later.[30] Most people have faith that very well-grounded results will find their way through the system. And it is right that doing so should be difficult. When I talked about this to the neuroscientist Paul Fletcher – whom we shall meet properly later – he pointed out: 'To an extent, science *has* to be robust to new things; it has to say: "I don't believe you yet." Otherwise it becomes incredibly flimsy, and it just flip-flops each way.' The friction, and even bullish resistance, of the publishing system, then, in theory help to ensure that only the most robust findings will pass muster. If this marketplace were more liquid and efficient, we'd be overwhelmed with speculative nonsense.

Even peremptory or aggressive dismissals of new findings have a crucial place in the intellectual ecology. It is all too easy to point and laugh at people in the past who resisted an idea that later became orthodoxy – Georges Cuvier's rejection of evolution, Ernst Mach's rejection of atoms – but in doing so they not only helped the new idea to be reformulated with ever greater persuasive power; they were actually conducting science in an impeccable fashion. As Thomas Kuhn points out, even 'lifelong resistance' to a revolutionary new idea 'is not a violation of scientific standards but an index to the nature of scientific research itself. The source of resistance is the assurance that the older paradigm will ultimately solve all its problems, that nature can be shoved into the box the paradigm provides. Inevitably, at times of revolution, that assurance seems stubborn and pigheaded, as indeed it sometimes becomes. But it is also something more. That same assurance is what makes normal or puzzle-solving science possible.'[31]

In other words, science would not be so robust a means of investigating the world if it eagerly embraced every shiny new idea that came along. It has to put on a stern face and say: 'Impress me.' Great ideas may well face a lot of necessary resistance, and take a long time to gain traction. And we wouldn't wish things to be otherwise. In this sense, the marketplace positively needs to be riddled with friction and inefficiency.

Place your bets

The American investor Warren Buffett, also known as the Wizard of Omaha, is an extremely rich man. Personally I also find him sympathetic because he hates meetings. You can't make an appointment to see Buffett more than twenty-four hours in advance. Want to see him? Call the previous day and he'll see if he's got time. It doesn't seem to have done him any harm. But of course that can't be the whole secret of how he got so rich. So how did he do it? By rethinking an old idea and running with it when almost everyone else thought it was stupid.

When Buffett was four years old, the Columbia business professors

Benjamin Graham and David Dodd published a book called *Security Analysis*. In it, they described the principles of what later became known as 'value investing'. To put it very simply, a value investor looks to buy stocks at less than their intrinsic value, so as to have a 'margin of safety' against market fluctuations, and holds them long-term.

No one in the business mainstream took much notice, but Warren Buffett later studied with Graham and moved into the investment world, formally taking control of his company Berkshire Hathaway in 1970. Along with a few other brave souls, Buffett followed the old-fashioned principles of value investing rigorously, even though right up until the 1990s it was little more than a joke to mainstream financial theorists. As Buffett later commented wryly: 'You couldn't advance in a finance department in this country unless you thought that the world was flat.'[32]

Value investing was eventually taken seriously, in large part thanks to Buffett's own successful example. You can't argue with a personal fortune of $65 billion. But if everyone had been investing according to the same theory all along, there would have been more competition and so Buffett couldn't have achieved such impressive margins. It can pay off handsomely to hold the minority view. It was crucial to the magnitude of Buffett's success that almost everyone else had a different theory. In the marketplace of ideas, he was a fringe player.

When you notice different valuations of securities or currencies across markets, and make a profit by trading on the difference, that's called arbitrage. So as well as taking advantage of different opinions about the market value of securities, Warren Buffett was profiting from arbitrage in the marketplace of ideas. Because few others at first subscribed to the idea of value investing, he was able to profit handsomely from it. His strategy was underpriced. Buffett was betting against both the market *and* the marketplace of ideas, and winning.

In their different ways, both Warren Buffett's financial success and the persistence of bizarre conspiracy theories demonstrate the inadequacy of any notion of an efficient marketplace of ideas. In this market there is no equilibrium, there are frequent crashes and failures, and – as we'll soon see – dysfunctional products *must* sometimes succeed. Plus, it's infested with zombies. Evidently, we can't trust the marketplace of ideas to do our sifting of ideas for us. Often we'll be right to bet against it, as Buffett bet against the market in investment theories. For many reasons, then, it's important to rethink the free-trade metaphor that the best ideas always rise to the top. The very notion that they do is itself a zombie idea that helps entrench powerful interests.

On the other hand, zombies can be useful when they motivate energetic refutations that advance public education. Yes, we may regret that people often turn to the past to renew an old theory like Flat Earthism that really should have stayed dead. But some conspiracies are real, and science is always engaged in trying to uncover the hidden powers behind what we see. The resurrection of zombie ideas as well as the stubborn rejection of promising new ideas can both be important mechanisms for the advancement of human understanding.

9

How to Be Wrong

But a wrong idea returning can be better than no idea at all. Being wrong can be useful, in reminding us of what we don't know.

'We are all agreed that your theory is crazy. The question which divides us is whether it is crazy enough to have a chance of being correct. My own feeling is that it is not crazy enough.' – Niels Bohr

In 1981, a brilliant plant biologist published a book that received respectful notices, until a review in *Nature* appeared with the headline 'A book for burning?' From then on, the author was *persona non grata* in scientific circles. Three decades later, he gave a talk based on his latest book in Whitechapel, London, as part of TEDx, the satellite franchise of the TED-Talk operation. After some heated complaints, TED took the video of the biologist's talk down from its main website and declared it unscientific. (Later they had to retract their declaration.) What is it about this guy that makes people so angry?

In person, Rupert Sheldrake is hardly a fire-breathing polemicist. Indeed, he is terribly polite, professorial, and soft-spoken, unthreateningly dressed in a blue jumper and oatmeal cords. He makes me a cup of tea in his Hampstead kitchen before we go upstairs to the study to talk. And yet, since his first book changed his life and career in the 1980s, he has become somewhat dangerous to know. He no longer works at any scientific institution. 'I'm a feral scientist,' he says with a twinkle. His ideas are considered taboo. But to declare any idea unthinkable is to risk missing out on something extraordinary.

What goes around

It's annoying when zombie ideas from the past return again and again. But what complicates matters is that often we should be ready to give wrong ideas a second chance. It's possible to be right in the wrong way – simply by guessing, but giving no reason for anyone to take you seriously. And conversely, it's possible to be wrong in the right way. Some wrong ideas are even *necessary* wrongs – the only possible stepping stone to being right in the future.

And this is the case simply because we are all too human. Just as orthodox economics assumes that consumers always make perfectly rational decisions in the marketplace of products, so we are urged to believe that people always make perfectly rational and sound decisions when choosing ideas in the marketplace. But that is just not true. People can be right for very unsound reasons. And wrong for very good ones.

Stories of scientific revolutions in the past often turn out to be stories about wrong theories forcing correct theories to raise their game. In the absence of such stimulation, the right theories might never have got thoroughly worked out. Such cases are usually thought of as ones in which market competition worked as advertised. This is indeed one common way of looking at, for example, what is called the 'Copernican revolution' in astronomy, the triumph of the theory that the Earth goes around the Sun, and not vice versa.

Amazingly, geocentrism – the opposite of Copernicanism – has itself been revived as a zombie in our own time. (In spring 2014, a web documentary proclaiming that the Sun really does go around the Earth, *The Principle*, featured apparently supportive interview clips with famous scientists, such as the physicists Lawrence Krauss and Michio Kaku, who later said they had been misled as to the nature of the film, and that their contributions had been 'cleverly edited'.)[1] But the story of how geocentrism was originally abandoned is more complicated than we are usually told. The popular narrative goes that in 1543 Copernicus published his book *De revolutionibus*, which announced – or rather re-announced, since the Greek philosopher

Aristarchus had considered this obvious two millennia previously – that the Earth went round the Sun; Galileo later checked with his telescope; and the scales quickly fell from everyone's eyes, a bit of religious persecution notwithstanding.

The problem with this story is that Copernicus was wrong.

The great Danish astronomer Tycho Brahe was a red-headed genius, born three years after the publication of Copernicus's book. At the age of nineteen, Brahe fought a duel over a quarrel about mathematics. His opponent managed to cut Brahe's nose off. Not at all discouraged from pursuing scientific disagreements, Brahe simply made himself a prosthetic nose out of brass and carried on with his learned pursuits.[2]

In 1572 Brahe noticed a new star in the heavens, brighter than any other. That wasn't supposed to happen. According to the prevailing tradition inherited from Aristotle and Ptolemy, the supralunary realm – beyond the Moon – was eternally fixed and unchanging. And yet, using measurements and simple trigonometry, Brahe was able to show that this new star really was out there, beyond the Moon. He called it a 'nova'. This was such an important discovery that the king of Denmark (whom Brahe's stepfather had once saved from drowning) gave Brahe an island and what he called a 'ton of gold' to build a world-beating astronomical observatory. There Brahe created new instruments and buried them in bunkers to protect them from the vibration of the wind.[3] He settled down to watch the heavens with his wife and dog, whom he named Lep the Oracle, and occasionally pretended to consult on difficult astronomical questions.

Brahe had discovered what we now call a supernova – an exploding star. (His theory that it was a new star is still there in the name: *nova* is Latin for new.) And yet he – like nearly all other thinkers of the time – didn't buy Copernicus's theory that the Earth went round the Sun. Why not? Because there were at this time many good reasons to reject Copernicus's theory – good *scientific* reasons.[4]

In 1588, Brahe responded to Copernicus's theory with his own ingenious 'geoheliocentric' model of the solar system, in which all the other planets did indeed go around the Sun (as in Copernicus), but the Sun-and-other-planets system itself orbited the Earth, which was still serenely placed at the centre of things. Brahe's new system was mathematically exactly equivalent to Copernicus's. So heliocentrism offered no advantage in predictive power over Brahe's version. What was worse, Copernicus required people to believe that the great hulking mass of the Earth moved, a highly counter-intuitive idea for which Copernicus himself could offer no good explanation. (What kind of enormous invisible power could move the whole world?)

Brahe pointed out some other problems, with alacrity and no lack of scientific rigour. One was the implied size of stars in Copernicus's system. If you look at a star in the night sky, it isn't an infinitely small dot – it has width. If you measure that width, and accept that the stars lie as far away as Copernicus said, then elementary geometry tells you that all the stars must be incredibly huge – far, far, bigger than our Sun. Brahe said that it was absurd to conceive of such gigantic stars. (The problem is cleared up by our modern knowledge of optics: the apparent width of stars when viewed with a telescope or the naked eye is an illusion created by the physics of how light waves pass through a lens.)

These controversies, in fact, ran long beyond Galileo's observation in 1611 of the phases of Venus, which some historians take to mark the point at which everyone converted to Copernicanism. (Like the Moon, Venus varies in appearance from 'new' or near-invisible, through slim and fatter crescents, to a complete orb. The compelling inference is that this is because of the varying amounts of light it reflects in our direction as it orbits the Sun. When Venus is between Earth and the Sun, it is 'new', or dark, because it is taking all the sunlight on the face turned away from us; when it is on the opposite side of the Sun, it is 'full', because the Sun's light is bouncing off its whole face back to Earth.) Yet half a century later, the Italian astronomer Giovanni Battista Riccioli pointed out that if the Earth really were spinning and speeding through space around the Sun, the path of falling objects near the Earth's surface ought to be slightly perturbed by this infernally rapid motion. But no such deviation could be observed. Riccioli was exactly right about this: the trouble was that instruments at the time were nowhere near precise enough to catch what, centuries later, was in fact measured and identified as the Coriolis effect.

These and other problems were, for a century and more, good reasons to be suspicious of Copernicus. They were scientific reasons – unlike those of Martin Luther, who called Copernicus a 'fool' comparable to Joshua in the Bible, who likewise vainly 'bid the sun to stand still and not the earth'. By 1674, Copernicanism had become the majority view among astronomers, yet Robert Hooke of the Royal Society still pointed out that no one had been able to demonstrate definitively its superiority to the 'Tychonic' system. According to some historians, indeed, for a long time Brahe's model 'fit the available data better than Copernicus's system did'.[5]

And in another, important sense it was *more* revolutionary. Copernicus's system retained the existence of the nested set of 'heavenly spheres' proposed by the ancients: indeed, his book was entitled *De revolutionibus orbium coelestium*, or 'On the orbits of the heavenly spheres'. According to Copernicus, each known planet (plus the Moon) had its own sphere, and the outermost sphere held the fixed stars. The spheres were invisible crystalline solids, and their rotation produced the orbits of the planets around the motionless Sun at the centre. When Tycho Brahe was calculating his own system, however, he noticed that it implied that comets crossed through the planetary spheres, which were supposed to be made of 'hard and impervious matter'.[6] What to do? Brahe trusted the maths and threw out the spheres. That was a real revolution.

Brahe decided that the planets simply floated freely in space. He also envisioned an infinite universe with numberless suns and numberless planets like Earth that might support extraterrestrial life. He said the universe had no centre (unlike Copernicus, who put our Sun at its centre), and that all motion within it was relative. He was, too, the first to insist that our solar system's planets shone solely by reflecting light from the Sun. For many reasons, Tycho Brahe's 1588 publication of his geoheliocentric system is now considered to mark the real birth of modern astronomy.[7] Yes, Copernicus was right on the main point – but for a long time, his critics were in many ways doing better science.

Eventually Copernicanism was accepted, but not because it was perfect. After Galileo's observations and some mathematical refinements that improved on Brahe's competing model, it was just more useful. (Like the idea of possible worlds in philosophy.) A major factor in its adoption, as it happens, was that it made the predictions of astrologers more accurate – another wrong reason to be right.

The right kind of resistance to a new idea, as the Copernicus story shows, is an essential component of what is the irreducibly social and collaborative process of human inquiry. Indeed, as Thomas Kuhn emphasises in his seminal *The Structure of Scientific Revolutions*, there has never been any such thing as a theory completely free of troubling anomalies and apparently contradictory evidence.[8] And the people who turn out to be wrong can be just as important as those who turn out to be right.

Against principles

Much later, Copernicus's name was borrowed for a guiding scientific assumption known as the Copernican Principle. Just as Copernicus himself had demoted Earth's place in the grand scheme of things by removing it from the centre of the solar system, so the Copernican Principle says that we – along with our solar system – don't inhabit any particularly special place in the universe as a whole. We just happen to find our unremarkable star in an unremarkable part of an unremarkable galaxy. What we see around us is a perfectly ordinary part of the universe. There's nothing privileged about where we live. This is a good working assumption – unless one day we find that it isn't.

Modern science is underlain by a lot of such assumptions, but they should always be open to question. A bold recognition of the provisional nature of scientific assumptions was a feature, for example, of the revolution in physics in the early twentieth century, with its bizarre ideas that subatomic phenomena could be both waves and particles at the same time, and its overarching notion that reality was probabilistic rather than strictly determined. Albert Einstein famously objected to the latter idea by asking the quantum pioneer Niels Bohr whether he really thought that God played dice. Bohr remembered later that Einstein 'expressed a feeling of disquietude as regards the apparent lack of firmly laid down principles for the explanation of nature', while Bohr himself thought that 'we could hardly trust in any accustomed principles' save that of logical consistency.[9]

Another scientific assumption widely shared today is that the laws of nature are the same across space and time. The strength of the gravitational force, or the rate of atomic decay of uranium – these things are the same at the other end of the universe as they are here, and billions of years ago they were the same as they are now. It sounds commonsensical. Which ought to be our first alarm bell. In fact, this too is just an assumption.

For a start, speaking in general philosophical terms, it's not clear what 'laws of nature' really are. The phrase is a metaphor taken from human society, where laws are written statutes that are obeyed or ignored as people choose. It is then applied to the cosmos, where subatomic particles and the rest have no choice but to obey these laws. Indeed, they *cannot* be ignored, on pain of their no longer being laws. But where do these 'laws' exist? In some Platonic realm? How does matter know how to follow them? (How easy it is to revive teleology and panpsychism at this point.)

The phrase 'laws of nature' originated in the seventeenth century, and it meant laws that God imposed upon reality. The chemist (and alchemist) Robert Boyle, for example, wrote: 'The Wisdome [. . .] of God does [. . .] confine the Creatures to the establish'd Laws of Nature.'[10] Isaac Newton discovered the laws of motion and gravitation that he believed had been set down at the beginning of time by the Creator.

Stripped of its theology in modern times, the idea of a law of nature becomes more, not less, mysterious. We never perceive laws of nature directly; instead we observe mathematical regularities in phenomena and infer general rules that fit these patterns. It might be that the laws are different in other universes within the multiverse. (Recall that this is the cosmological argument put forward to answer the fine-tuning problem.) And if laws of nature can differ across universes, who is to say that they cannot differ within them, undergoing perhaps some kind of evolution through time?

This possibility has been raised before. It was suggested by two celebrated physicists: in the 1930s by Paul Dirac, and again in the 1970s by John Wheeler. It has arisen again in our time in the work of the physicist John Webb, who has suggested that the laws of nature we observe might just be the 'local by-laws' of our part of the universe, and that supposed 'constants' might themselves have changed over the life of the universe.[11] Yet to say that the laws of nature might change is still, in mainstream discussion, the kind of thing that can get you branded a crank, or worse.

Paradigm busting

Rupert Sheldrake could have foreseen he was going to annoy people by entitling one of his books *The Science Delusion*, which was also the title of his now-notorious TEDx talk. But it is not an anti-science screed, in the way that Richard Dawkins's *The God Delusion* is an anti-religion screed. (His British publisher, he says, insisted on the title; the book is published in the US as *Science Set Free.*) The 'science delusion', for Sheldrake, is merely the erroneous belief that science already has a good understanding of the nature of reality. His book and talk point out, affably and in fine historical detail, why many of the ruling scientific assumptions of our day are just that – assumptions. That there are eternal laws of nature; that matter is unconscious (Sheldrake, like Galen Strawson, is inclined to panpsychism) – such propositions have been adopted as working hypotheses (often for very good reason), but, Sheldrake argues, they should not harden into dogma. They may have to be rethought.

So far, so reasonable. The TEDx event Sheldrake spoke at, he points out, was called 'Challenging Existing Paradigms', which is exactly what he was doing. Yet the people who demanded that the video of his talk be taken down seemed to feel that any weakening of the homogeneous truth-edifice of science is impermissible, since it might open the door to irrational superstition. These people included the highly respected evolutionary biologists Jerry Coyne and P. Z. Myers, who have done excellent work in attacking the political promotion of creationism in US schools. (It is probably this context that explains their particular hostility. Most of the prominent scientific atheists of our time are biologists; you don't hear so many public attacks on religion from physicists or cosmologists.) But Sheldrake's own material wasn't offensive to secularists on these grounds. 'I mean the whole of my theory is evolutionary,' he says. 'So it's not as if my position is remotely linked to creationism.'

In *The Science Delusion*, however, Sheldrake was not simply politely pointing out how science rests on assumptions. He was offering – equally politely – some positive suggestions of his own, his own 'evolutionary' theory. It encompasses the idea that the laws of nature might evolve, but a lot more as well. This idea, which he calls morphic resonance, is what has got him into trouble over the years. Even though it is not completely new either.

Instead of thinking of nature as law-governed, Sheldrake suggests, why not think of it as habitual, as operating through memory? This, he is the first to point out, is not an original thought. He had it while working as a biologist as a fellow of Clare College, Cambridge, in 1973 – and he had it because others had had it before. 'I got the idea,' he explains, 'partly through looking at ideas that had been dismissed and completely rejected.' We've seen, so far, how that can be a recipe for triumph in fields from science to chess. But in this case the story is more complicated. It might be a case of misguided rethinking. But one thing Sheldrake's experience shows for sure is the human – and perhaps also scientific – cost incurred when the establishment decides that reviving an old idea is simply impermissible.

*

So what did Sheldrake rethink? He looked again at a historical concept in the science of biology that today is deader than the dodo – vitalism. According to a vitalist, there is something special about living things, and they cannot be totally explained by reference to the physical interactions of matter. Today, the term vitalism is a byword for historical superstition. It is taboo. When Rupert Sheldrake was a student, he remembers, 'Biology textbooks would start, "People used to think there was some special 'vital force' that organized [living things] *but we now know* it's nothing but regular physical and chemical mechanisms, et cetera."' Vitalists were the 'baddies', he says: 'If there was a scientific pantomime, the vitalist would have had a green light when he came on stage, and everyone would have hissed. But nobody actually read them; I sometimes asked people, my supervisors and scientific colleagues and friends, "Well what do you think the vitalists actually said?" And they said "Oh, they just think there's a *vital force* . . . a useless concept."'

In his own research, however, Sheldrake had decided that 'mechanistic biology was heading towards a brick wall', and that genes were 'hopelessly overrated' as a way to explain the inheritance of instinct and biological form. In general, he recalls, 'The attempt to reduce life to molecules and molecular interactions just seemed to me *so far* from being able to understand things like homing pigeons, or migration, or consciousness . . . and principally the development of form, which I was working on. I was working on plant morphogenesis. How do plants grow, how do animal embryos take up their shape?' At the time this seemed very difficult to explain mechanistically. So he thought: 'Well, maybe it's worth looking at what the vitalists actually said.'

Looking into the vitalists, Sheldrake was especially intrigued by the work of Hans Driesch. 'He was a very brilliant embryologist,' Sheldrake says, 'and he pointed out that no amount of analysis in terms of genes or molecules – anticipating the way things were going – would ever solve the problem, because you have to have some kind of top-down, goal-directed process to explain living organisms.' (Most modern evolutionary biologists would strongly disagree, confident that the explanatory power of bottom-up explanations, such as genetics, will continue far into the future.) At the time Driesch attempted to revive a kind of Aristotelian teleology, but it was no more respectable then than it is now. Later, scientists changed the language, proposing in the 1920s the idea of 'morphogenetic fields'. Analogous to an electromagnetic field, a morphogenetic field somehow directs the growth of what is in it, whether living or not. This marks a shift from a vitalist model to an 'organismic' model – which says that not just living things but *all* things require some sort of top-down, governing force to organise them. Rupert Sheldrake agreed, so he became a follower of organism. 'Whereas most people thought vitalism went too far,' he notes wryly, 'I thought it didn't go far enough.'

The problem that Sheldrake was wrestling with in 1973 was this: if morphogenetic fields really exist, how can they be inherited? Following the philosopher Henri Bergson's unusual ideas about how human memory worked, Sheldrake decided that perhaps morphogenetic fields could be 'inherited, as it were, non-materially: by a direct connection across time'. This process he christened 'morphic resonance'. He thought at first that it should be purely a biological principle. But then, 'My friends said to me: "Well, what about crystallisation?"' At the time it was impossible to predict the precise form any compound would take when it crystallised, which is an important problem in the pharmaceutical industry among others. Sheldrake thought that a related problem might be protein-folding, the process by which a chain of amino acids assumes its final three-dimensional shape. (There is a ridiculous number of possible combinations.) According to Sheldrake's 'morphic resonance' idea, the formation of crystals and proteins, as well as that of plants and animals, would be driven by habit owing to the connection of immaterial 'morphic fields' across time. As he recalls thinking: 'This is a much more radical revolution than I thought. And is much more sweeping in its implications.'

Sheldrake, who was now living in India, decided to write a book. He was warned it would be controversial. 'Friends in Cambridge said: "Don't publish this theory, because you'll wreck your career",' he recalls. 'You know, I got a double first, I got a research fellowship, I'd become a fellow of Clare, I was director of studies, I had a Royal Society research fellowship, I had paper after paper in *Nature* and other top journals . . . And so it was quite a big decision. I realised how intolerant the scientific world is. People at Cambridge weren't hostile, they just thought this was perverse – why would I throw away a promising scientific career to pursue some madcap idea that seemed to them entirely improbable?'

He had to feel confident. So, after four years thinking about it, Sheldrake discussed morphic resonance with some Indian colleagues, and found out that the idea of a memory in nature is mainstream in Hindu and Buddhist traditions. 'So something like a large proportion of humanity had been thinking along those lines for many centuries if not millennia,' he says. 'In a sense that was reassuring – it's not just individual madness on my part. Here's a long philosophical tradition, and these people seem perfectly sane.'

Eventually, in 1981, Sheldrake's book was published. It was called *A New Science of Life*. Doesn't that sound like rather a hubristic title? 'Well, yes,' Sheldrake says good-humouredly. In fact his original title was *Towards a New Science of Life*. But his publisher, Anthony Blond, wanted to remove the modest, hesitant 'towards'. Sheldrake remembers: 'I said, "Well, that sounds as if it's rather exaggerating the claim, I'm talking about something that's *towards* a new science of life." Blond said, "Trust me, dear boy, I published Schumacher's book *Small Is Beautiful*. I made up the title and that's why it sold. The original manuscript was called *Some Aspects of Small-Scale Economic Planning*." Sheldrake laughs. 'So he said, "I'm a publisher: this is what I do. Do you want the book to sell or don't you?" And I said, "Yes." He said, "Well then, its title is *A New Science of Life*."'

Three months after the book's publication, the review in *Nature* appeared. The journal's editor, John Maddox, wrote that 'this infuriating tract' was 'an exercise in pseudo-science', and already 'well on the way to being a point of reference for the motley crew of creationists, anti-reductionists, neo-Lamarckians and the rest'. It was, Maddox said, 'the best candidate for burning there has been for many years', but since burning books was not after all a good idea it should instead just be 'put firmly in its place among the literature of intellectual aberrations'.[12] Sheldrake's career was destroyed, and he never worked in a scientific institution again. In 1995, John Maddox was knighted.

Back in the field

Scientific orthodoxy isn't necessarily suspect just because it's orthodoxy. The establishment is usually right: that's why it's the establishment. And the conventional wisdom is usually true: that's why it has become conventional. (Though no one calls it 'conventional' unless they are announcing they are about to overturn it.) So the conventional wisdom, for example, that general relativity gives a pretty reliable account of accelerating massive objects in space is the reason why the GPS in your phone works. Anyone proposing a new idea that seeks to overturn that conventional wisdom, but fails to preserve its predictive power, will with justice be dismissed as an amateur or a crank. And so too it might very well be that morphic resonance is widely rejected today not because institutional scientists constitute a defensive priesthood but because few people see any very compelling reason to explore it further. Not all old ideas are worthy of revival.

Sheldrake claims that morphic resonance explains an awful lot of things, which go far beyond plant growth and crystal formation. Analysing the results of some experiments in rat learning conducted during the first half of the twentieth century, he suggests controversially that when rats learn their way through a maze in one city, rats all over the world should subsequently learn the same maze more quickly, via the shared global rat memory.[13] Indeed, his most successful book to date is about 'animal telepathy', which supposedly becomes possible when animals tune in to morphic fields. That book, *Dogs That Know When Their Owners Are Coming Home,* has sold half a million copies in the US alone.

Sheldrake's interest in telepathy is enough for some to refuse to take him seriously. In fact, for a decade and a half after the initial publication of *A New Science of Life*, he had tried to get 'hard' morphic-resonance experiments done, suggesting experiments on crystallisation to friends in industrial chemistry laboratories and on fruit flies in university biology departments. (One prediction of morphic resonance, for example, is that repeated crystallisation of a compound ought to make the melting point of the crystals rise over time.) But all such efforts were shut down because of Sheldrake's reputation as someone who wasn't taken seriously by the establishment. His experiments never got done. So, lacking a laboratory, he eventually decided to do the only experiments he could, cheap ones: pets waiting at windows; human beings sensing that someone is staring at them (again by tuning in to morphic fields). 'I mean, at last,' he says, 'I could actually do something scientifically. My mind works along the lines of thinking up experiments – I almost *dream* of experimental designs – and here at last I could actually get involved in doing them.' The results of those studies, in journal articles and Sheldrake's books, are controversial – others have disagreed with the experimental designs or Sheldrake's interpretation of the data – but it is hard to say that they do not constitute scientific inquiry. And the subject is one that an awful lot of people are interested in.

Most dog owners, Sheldrake points out, think that their pets know when they're coming home. 'And the sense of being stared at,' he says, 'over ninety per cent of people have had that experience. "Yes, that happens to me," they say. Telephone telepathy?' – the sense of knowing when someone is going to call you – 'It's over eighty per cent. These are normal, everyday experiences for a lot of people. And it seems to me an extraordinary fact that the entire institutional science structure is in denial about these things, even though most individual scientists have had the experiences themselves.' Such common experiences, indeed, are generally assumed to be illusory. They can, it is usually said, be explained away by our psychological quirks such as confirmation bias, or by unforeseen experimenter effects, in which researchers unconsciously influence subjects during formal studies.

Maybe indeed there is nothing to them, Sheldrake says. But why not find out?

If telepathy is possible, of course, it would be the mother of all black boxes. But that alone is no reason to rule it out. Yet the mainstream psychology community is not just sensibly sceptical about such phenomena, generally called 'psi'. It complains bitterly when anyone takes them at all seriously. In 2011, Daryl Bem, an emeritus psychologist at Cornell, published a paper in a top journal detailing apparently positive results with experiments on precognition. The paper, entitled 'Feeling the Future', implied that subjects' present behaviour could be influenced by future events.[14] People could remember words more easily if, *later*, they were asked to repeat them several times. In other words, causation was working backwards in time. These effects were small but, Bem argued, significant. Well, you might think, that is extremely surprising, and interesting if true. The journal's editor, psychologist Charles Judd, said that the paper had been accepted because it met their editorial and scientific standards – even though, he added, 'there was no mechanism by which we could understand the results'. (He thus showed himself prepared to countenance the existence of black boxes.)[15] But other scientists queued up rapidly to denounce him: the experiments were 'a waste of time', one said.[16] 'It's craziness, pure craziness,' said another, 'an embarrassment for the entire field.'[17] According to one survey, in fact, around a third of psychologists consider psi to be impossible. Not unlikely; impossible. Later, Bem published with colleagues a 'meta-analysis', statistically reanalysing the results of ninety studies on psi phenomena: the paper claimed again to see positive effects.[18] Others responded that, if so, the standard methods of statistical analysis used throughout the field must themselves be flawed.[19] (Perhaps so.) Bem himself said he had always been a 'maverick', but if he had been a young researcher trying to make a career, such research would have been fatal for him. He could only get away with it once retired from a highly respectable research career, and even then he was greeted with the dismay and opprobrium of his colleagues. But the larger point is that, whether Bem is right or wrong, the insistence that any such phenomena are literally *impossible* must be accounted an unscientific one.

Rupert Sheldrake proposes, modestly, that one per cent of the public science budget might be devoted to research suggested by laypeople. A lot of that research, it seems, would involve animals. 'In Britain,' he points out, 'something like forty per cent of households keep pets . . . there's an enormous interest in dogs and cats and domestic animals; horses and so on. That's why newspapers, whenever there's a story about a cat that finds its way home from miles away, they love these stories. Journalists love this stuff. Videos on animals go viral on YouTube all the time: people are really interested in this. And yet, if you go to institutions of science, then the research on animal behaviour is very restricted, on the whole.'

Come to think of it – how *does* a cat or dog find its way home from miles away? The truth is, we don't know. And Sheldrake's general point that research could spread its net wider – so as to encompass things that might turn out to be nonsense yet might reveal surprising things we hadn't foreseen – is one shared by many respectable thinkers. A similar argument is made, for example, by the historian of technology David Edgerton, who argues that countries should not imitate one another's 'innovation policies'. Why not? 'For if all nations, areas and firms are agreed about what the research should be, by definition it will no longer be innovative; and it might not be a good thing that all nations pursue the same policies for research, because they are likely to come up with similar inventions.'[20]

This is so because innovative research, by definition, is research that everyone doesn't already agree is the right kind of research to pursue. The British biochemist Peter Mitchell, for example, was widely considered an outsider in the 1960s, doing obscure experiments in his own private laboratory in Cornwall. At the time, Sheldrake says, he was 'a complete heretic!' But Mitchell revolutionised the understanding of how mitochondria in organisms create ATP, the molecular basis of energy transport. In 1978 he was awarded the Nobel Prize in Chemistry.

Of course, just because we can point to one maverick who was vindicated, it doesn't mean that all mavericks are right. One might well feel suspicious of morphic resonance, for example, on the grounds that it seems like a God-of-the-gaps idea. Here are all these things we currently don't understand properly; and here is one overriding concept that seems to unify and explain them. Some of those gaps, moreover, seem to have narrowed if not closed in modern times. Sheldrake claimed that the formation of crystals was completely unpredictable, but modern mathematical modelling techniques run on supercomputers have made inroads into the problem, as they also have in the field of protein-folding. The 'mechanistic' approach has not hit a total dead end yet.

Meanwhile, the extraordinary developments in genetics – mapping the entire human genome a mere half-century after Crick and Watson first hit on the right molecular structure of DNA – are hugely impressive. In a way, the continuing belief in vitalism well into the twentieth century was part of what spurred them: Crick himself declared he was devoted to stamping out vitalism.[21] As biology uncovered more and more of the physical and chemical processes that underlie the function and development of living organisms, more and more of vitalism was simply debunked. On the other hand, the reports from epigenetics today complicate the old story of inheritance. And Sheldrake had all along nurtured a Lamarckian conviction that acquired behaviours could be inherited, in order for instincts to evolve. The evolutionary biologist Stephen Jay Gould, in his magisterial *Ontogeny and Phylogeny*, describes the common nineteenth-century Lamarckian view like this: 'Instincts,' people believed, 'are the unconscious remembrance of things learned so strongly, impressed so indelibly into memory, that the germ cells themselves are affected and pass the trait to future generations.'[22] That idea, as we have seen, became taboo for most of the twentieth century. Yet in 2013 it was reported that mice who had been conditioned to fear the smell of a cherry-like chemical had offspring who, from birth, also instinctively feared that smell.[23] Score one for Lamarck and Sheldrake.

More surprisingly, the idea of the 'morphogenetic field' itself has also stealthily crept back in to evo-devo, or evolutionary-developmental biology, the school that arose towards the end of the twentieth century out of a desire to bridge just that gap between genetics and development that had frustrated Rupert Sheldrake in the 1970s. The modern version of this idea is that a morphogenetic field controls whether a particular bunch of embryonic cells will develop, say, into a leg or an arm. Some biologists say that this means no more than that the area of cells is 'destined to form these particular structures' – though that does rather raise the spectre of teleology again.[24] Others identify the morphogenetic field with a particular physical structure or process: specific sets of interactions between genes and proteins, or, more abstractly, the distribution in space of chemical signals.[25] (These, it is said, encode the 'informational content' of the field that directs the organism's development.) Other researchers point to the fact that the behaviour of a morphogenetic field can be the same even when the underlying gene- and cell-related mechanisms are different. So, for example, the biologist Ellen Larsen suggests: 'The morphogenetic field has emergent properties, independent of the particular molecular entities which carry out the behavior of the field.'[26]

Emergent properties of something, you'll recall, are properties over and above the properties of that thing's constituent parts. Like the emergence of liquidity from water molecules, or the alleged emergence of consciousness from brain activity. Emergence, in this context, is the opposite of a 'reductive' explanation: one that focuses only on the smallest-scale mechanisms, ruled by physics and chemistry. The modern field of 'systems biology', too, studies dynamical and complex living systems with a holistic approach that is opposed to reductionism. Now, neither emergence in evo-devo nor systems biology postulate some kind of *immaterial* thing like Sheldrake's morphic fields – unless, that is, one considers that the concept of 'information' stealthily smuggles in a Neoplatonist element to such discussions. But the idea that there is something about living beings that cannot quite be understood at the smallest scales has not gone away. For emergence, as for vitalism, the whole has properties that the parts do not. Even at the microscopic scale, it turns out, many biological systems are not 'machine-like' in the way they operate: they work statistically or probabilistically rather than through mechanical causation.[27] Some biologists, therefore, argue that 'emergence' can be thought of as a newly respectable modern version of vitalism or organicism.[28] So even those old ideas may not be quite dead after all.

'The positive evidence for morphic resonance is not that strong,' Sheldrake happily admits, 'because there's been very, very little research on the subject.' (For this reason he modestly calls it a 'hypothesis' rather than a 'theory'.) Sheldrake's critics disagree strongly with him on the correct analysis of the data that his experiments have so far given. Yet that, too, is a scientific disagreement. They say his theory is unfalsifiable, but – as we have seen in cosmology – that is no longer so widely considered a killer takedown of scientific hypotheses. Sheldrake's idea at least makes testable predictions – those never-performed crystal experiments – whatever you think of its likelihood of truth. He has been called a heretic and much worse, but if he is wrong, he is wrong in the right way – as a scientist.

Stepping stones

Casting ideas into the wilderness as untouchable, and excommunicating those who dare to voice taboo notions, is not just unkind but intellectually myopic. History implies strongly that if we don't investigate things that seem unlikely, we may miss out on surprising truths. And we have seen many examples of such ideas coming back after a long time dead. There are more unknowns than we like to think. What's more, excommunicating ideas is foolish even if they turn out to be wrong. For the history of science shows that wrong ideas can be crucial. Copernicus's heliocentrism was wrong in that it preserved the heavenly spheres; Tycho Brahe's rival theory was right to destroy them. Progress in our understanding can require temporary adherence to scientific theories that are imaginative probes destined to crash-land into an inconvenient asteroid and fall apart. Sometimes you just have to be wrong first to be right later on.

The idea of 'dark energy', for example, is a currently fashionable topic in cosmology. But don't bet the house on it yet. 'When we look back in fifty years' time,' the cosmologist Andrew Pontzen muses, 'we may think that the idea of dark energy was terribly naïve.' That is surprising, because dark energy is not only a cutting-edge scientific probe, it's a spectacular example of rethinking. Although this time the person posthumously vindicated is someone who you might think needed no rehabilitation whatsoever. His name was Albert Einstein.

Einstein got a lot of amazing things right. But there was one important thing he got wrong. In his initial formulation of general relativity the equations included what he called a 'cosmological constant', which described the energy density of the vacuum of space. Later, Einstein thought he had made a mistake and regretted not having reached into the equations manually to set his cosmological constant to a value of zero. (It is often said that Einstein called this his 'greatest blunder', though the evidence for his having said these words is at best second-hand.)[29]

But then the cosmological constant came back, with a vengeance. The problem is that the expansion of the universe is observed to be accelerating, and so there seems to be some kind of repulsive force acting against gravity, blowing everything apart. Scientists now call this dark energy, and it vindicates Einstein's original description of the cosmological constant to a magnificent extent. (The relevant numbers are the same to a high degree of accuracy.)

Just because Einstein has triumphed from beyond the grave, however, doesn't mean we now understand perfectly what is going on. After all, dark energy is called dark energy because no one knows what it is yet. It has never been directly measured. And, though it is the basis of countless scientific articles over the past couple of decades, it might turn out to be plain wrong. Maybe there is no such energy at all.

'We tend to think about this in quite a specific way,' Pontzen points out. 'We call it "dark energy", and we write down mathematical models to describe dark energy that fall into quite a small number of possible patterns. But it could well be the case that what's really going on – well, we just haven't dreamt it up yet. It could be that the reason the expansion of the universe is accelerating is, for example, that we don't properly understand the way that gravity works on such incredibly large scales, rather than that you need to put something extra in the universe, this thing that we call dark energy.'

'Right at the moment we can't distinguish it at all from Einstein's cosmological constant,' Pontzen says. 'So it's possible that all we're seeing is that the cosmological constant is not zero, but then people ask: "Well, why would it be, like, 10^{-120}" – some tiny, tiny number? You'd think either it would be 1, or 0. It seems so close to zero it doesn't seem quite to make sense. So I think that's why people are looking for some kind of deeper explanation than just "It's a number". But what form that explanation finally takes – who knows?

'A lot of people are aware that we may be being a bit blinkered about the way that we think about dark energy,' Pontzen concludes, 'but on the other hand, what are you supposed to do about that? If that's your best avenue to explore, you've got to do it.'

It might be right for science to be impressed by something that turns out to be wrong – if it looks like the only game in town, the single stepping stone that leads us onwards. Some popular histories of science tend to mock refuted old ideas such as phlogiston (the name of a purported fiery element that was released from burning things) or the luminiferous ether – the universe-filling substance that provided a medium through which light waves could propagate. But we *had* to believe in the ether first, in order to get to where we are now. 'The idea of the ether was a necessary stepping stone,' Andrew Pontzen explains, 'and it inspired experiments like the Michelson–Morley experiment, which were these crucial experiments showing that the predictions that the ether was making were wrong.' In the late nineteenth century, the physicists Albert Michelson and Edward Morley reasoned that the ether had to be in motion relative to Earth, because of our planet's motion around the Sun. So in 1887 they conducted a brilliant, sensitive experiment to measure the speed of light in various directions. It had to be travelling faster, after all, when the ether was going in the same direction. But no such difference in speed was found – an amazing result at the time, which led directly to Albert Einstein's formulation of special relativity a couple of decades later.

So the ether, Pontzen points out, is 'a classic example, really, where a wrong idea is necessary to get to the right idea. It's very easy to judge things too harshly, and we probably shouldn't.' As the physicist Frank Wilczek puts it, indeed, in a variation on David Lewis's appeal to usefulness, the 'fertility' of an idea – whether it prompts further research and rethinking – is often more important, in the history of science, than its truth.[30]

Science, and human endeavour in general, needs its heroes of wrongness just as much as its heroes of rightness. Not just because the former inspire and provoke the latter; but also because few inquirers are ever *totally* wrong or totally right. Tycho Brahe was wrong on the main point (what goes around what), but right about discarding the heavenly spheres. And Isaac Newton was right about an awful lot, but still a little bit wrong. (You need what supplanted his laws of motion – Einsteinian relativity – to get GPS to work properly.) What's more alarming is that, in the long run, we're probably wrong about everything.

Wrong again

In 1875, a German university student called Max Planck was told by his teacher not to go into physics because 'in this field, almost everything is already discovered, and all that remains is to fill a few unimportant holes'.[31] Planck ignored this advice, and in 1900 presented a paper on radiation that ushered in the revolution of quantum physics. His teacher had been wrong. Who is the equivalent of Planck's teacher today?

There's a powerful and even frightening idea – actually, an idea about ideas, a meta-idea – in the philosophy of science, which is known as 'pessimistic meta-induction'. Induction, remember, is reasoning from what has happened to what will happen. The Sun has always risen in the morning; therefore it will also rise tomorrow. So meta-induction is reasoning about how that kind of reasoning will go in the future. Should we always be confident that the Sun will rise the next day?

For science, it doesn't look good. Because every major scientific theory in human history has been wrong. The story of science is one of old theories – that illness is due to an imbalance of the four bodily humours, or that the Sun orbits the Earth – being overturned. Max Planck's lecturer couldn't possibly have known that, a quarter of a century later, his student would propose an outrageous idea – that energy comes in irreducible packets, or quanta – that was to revolutionise physics. So what makes us think our current theories are correct? It would be a pretty wild coincidence if you and I were living at the first moment in human history when a scientific theory was around that would *never* be abandoned in favour of something better. That has never been the case before. Why should it be now? Hence, *pessimistic meta-induction*. The inference we should draw for the future about our ability to draw inferences for the future ought to be pessimistic. We're basically always going to be wrong, and we're wrong right now.

Naturally, this is quite an old observation. Montaigne expresses it this way: 'Whenever some new doctrine is offered to us we have good cause for distrusting it and for reflecting that the contrary was in fashion before that was produced; it was overturned by this later one, but some third discovery may overturn that too, one day.'[32] Note that this idea cuts two ways. On the one hand it vindicates those struggling to overturn a scientific consensus – as Isabelle Mansuy and other epigeneticists struggle to change the genome-centred picture in biology. On the other hand it doesn't hold out much hope that they are permanently right in turn.

It is tempting to resist this idea and plead that pessimistic meta-induction is too, well, pessimistic. For one thing, even if our current theories are still wrong, they are usually broader explanations – and more reliable as predictive tools – than what they replaced. (Einsteinian relativity reaches the parts that Newtonian mechanics cannot reach, in situations of high speeds and strong gravity, but Newton's theories are still good enough for most everyday purposes.) And if we can't be sure that we are right, a consistent pessimist should also point out that we can't always be sure that we were wrong in the past. The history of human thought is not a linear sequence of false theories being replaced by true ones. It's more like a dark and tangled web, in which flashes of inspiration can be smothered and obscured for centuries, until someone finds them again and blows gently on the sputtering flames.

In the mean time, our current theories – even if they're wrong – are still demonstrably good enough to enable spectacular advances in medicine and technology. Being wrong is nothing to be scared of. Indeed, ideas we recognise as wrong can be a great help, in reminding us of how much we don't yet know.

Curiouser

Perhaps it is true, as Rupert Sheldrake and others think, that the laws of nature evolve. What we can definitely say is that the evolution of scientific ideas is itself a gradual process with no goal or endpoint at which we may hope to arrive at perfect truth.[33] It is a story of unending rethinking. And it depends on a subtle balance between knowledge of what we do know (provisionally) and knowledge of what we don't. It must keep an open mind – but not so open, as the old joke goes, that everything falls out. So how do we strike the right balance and give ourselves the best chance of recognising more mere assumptions and finding more known unknowns?

On the one hand, it is good to be ignorant. 'Thoroughly conscious ignorance,' said James Clerk Maxwell, 'is the prelude to every real advance in science.' 'You can be ignorant, too,' the biologist Stuart Firestein reassures prospective scientists. 'Want to be on the cutting edge? Well, it's all, or mostly, ignorance out there. Forget the answers, work on the questions.'[34] When he invites guest scientists to come and speak at Columbia University, he says, they 'come and tell us about what they would like to know, what they think is critical to know, how they might get to know it, what will happen if they do find this or that thing out, what might happen if they don't. About what could be known, what might be impossible to know, what they didn't know 10 or 20 years ago and know now, or still don't know.'[35] As Jochen Runde might put it, science – and human intellectual exploration more generally – is in the business of converting unknown unknowns into known unknowns, and then (with a bit of luck) into known knowns.

It's good to be ignorant – but, of course, not too ignorant. We normally think of curiosity as a desire to know something. Yet, paradoxically, we need to know something *before* we can be curious. We need to know what we don't know. That is the argument of a celebrated paper on the psychology of curiosity by George Loewenstein. He suggests that curiosity arises from a desire to close an information gap. And we cannot be aware of a gap unless we already have the information that surrounds it.

This itself was not a new idea when Loewenstein proposed it in 1994. Like so much else, as we'll shortly see, it was first suggested by William James a century before. Loewenstein explains: 'James proposed that "scientific curiosity" – the type of curiosity that most closely corresponds to specific epistemic curiosity – arises from "an inconsistency or a gap in [. . .] knowledge, just as the musical brain responds to a discord in what it hears".' Loewenstein then explains how his theory hews close to this idea of James's. 'Consistent with this view, the information gap theory views curiosity as arising when attention becomes focused on a gap in one's knowledge. Such information gaps produce the feeling of deprivation labeled *curiosity*. The curious individual is motivated to obtain the missing information to reduce or eliminate the feeling of deprivation.'[36] If true, this model of curiosity has the implication, as Loewenstein points out, that 'a failure to appreciate what one does not know would constitute an absolute barrier to curiosity'.

The bad news, Loewenstein continues, is that 'There is good reason to believe that such barriers' to curiosity 'are pervasive'. There is, for example, a well-known overconfidence phenomenon in decision theory, 'whereby people underestimate the magnitude of gaps in their knowledge', and it seems that people 'generally believe they have much more information about a topic than they actually do'.[37]

So if we value curiosity and want to encourage it, what to do? It seems that, for one thing, the 'Socratic method' of questioning has deep value in uncovering a person's assumptions and knowledge-gaps. Indeed, Loewenstein points out that for this reason it is probably better as a teaching method than 'simply encouraging students to ask [their own] questions' – because if they lack enough information to be aware of a gap, then they won't be curious and will need to be given more information first to 'prime the pump' of curiosity.[38] Then, Socratic questioning by the teacher can 'make students aware of manageable gaps in their knowledge'. 'The importance of knowing what one does not know may explain the success of the "Socratic method" of teaching,' Loewenstein concludes – and it also contributes to the success of the rethinkers we have met throughout this book, who knew that they *didn't* know for sure that an old idea was wrong.

Another productive way to overcome such hurdles of overconfidence and over-ignorance might be, to put it in Jochen Runde's language, to try to turn unknown unknowns – which we can't even be curious about – into known unknowns, by whatever means necessary. To exercise one's imagination about hypothetical scenarios; or to imagine that some accepted view is not right but a necessary stepping stone to something better. So our response to people like Rupert Sheldrake, who polemically remind us what we don't know, shouldn't be outraged defensiveness; it should be gratitude.

Innovation, after all, is especially difficult when you are dealing with unknown unknowns. 'It's such a difficult target, you know,' Andrew Pontzen says of trying to figure out whether dark energy is real. 'You're trying to come up with something you don't even know – you don't really know what the thing is you're trying to come up with. So all you can do is sort of plug away at specific problems and hope that at some point everything kind of comes together, then probably looking back we'll go "God this was ridiculous", like thinking about the ether now, it just seems so ridiculous and naïve. Well, when we look back in fifty years' time we may think that the idea of dark energy was terrible . . . Because if you have *no* idea, then what are you going to do? At least if you've got a wrong idea, then you can do something about it!'

In 1975, the Austrian philosopher Paul Feyerabend caused a counter-cultural fuss with the publication of his brilliant book *Against Method*. It showed that the popular picture of one unified 'scientific method' – which dragged mankind out of the age of superstition and into a shiny new era of truth – simply failed as a description of how science has actually proceeded through history. For this he was charged in some quarters with infecting impressionable minds with a noxious relativism, if not nihilism. Certainly Feyerabend aggressively attacked Karl Popper and the uselessness of any 'strict principle of falsification', which in his view 'would wipe out science as we know it and would never have permitted it to start'.[39] Yet overall, Feyerabend was rather cheerful about his story – particularly about the potential fruitfulness of error. 'There is no idea,' he wrote, 'however ancient and absurd, that is not capable of improving our knowledge.'[40] Indeed, he added: 'There is hardly any idea that is totally without merit and that might not also become the starting point of concentrated effort.'[41] This wasn't a glib anything-goes attitude: it was a reasoned conclusion from the history of scientific investigation. And as we have seen, wrong ideas can serve as important probes and provocations, helping thinkers refine their positions in a way that potentially wouldn't happen if their notions were instantly accepted in the first place.

Rupert Sheldrake may not be right to revive a form of vitalism in his morphic fields, but his idea that the laws of nature might evolve is one that has had highly respectable adherents in the past. Even his interest in the possibility of telepathy is an idea that periodically recurs, seemingly ineradicably, in academic psychology. These reformulations of old ideas might be wrong, but they perform a valuable service by pointing out where our knowledge is not as robust as we like to think.

And in the mean time, wrongness can be very productive. The idea that there are eternal laws of nature might turn out to be wrong – but it has so far enabled the stupendous achievements of modern science. Dark energy may not be the correct way to revive Einstein's cosmological constant, but it might be a necessary stepping stone on the way to something better, eventually. It is often better to be provocatively wrong, like Tycho Brahe's revolutionary but inaccurate picture of the solar system, than to be unsurprisingly right.

10

The Placebo Effect

And some old ideas are so powerful, it doesn't even matter whether they are true or not.

'The falseness of an opinion is not for us any objection to it.'
– Friedrich Nietzsche

All in the mind

This is the age of the brain. And thanks to the hi-tech science of fMRI scanning, we know a lot more about how it works than ever before. We know that the brains of people with mental disorders work differently from the brains of people without them. We know that they will benefit only from the most up-to-the-minute psychological techniques. More generally, thanks to all the triumphs of technologically powered medicine, we know that people should be treated with precisely engineered medications rather than being fobbed off with placebos.

Yet none of those things is true.

Which starts to become apparent after, one summer morning, I arrive at the Herchel Smith Building for Brain and Mind Sciences in Cambridge. It's a red-brick building in a leafy enclave within the sprawling grounds of Addenbrooke's, one of the UK's major research hospitals. Not long after I am buzzed in, a boyish neuroscientist bounds into the lobby and greets me. He is Paul Fletcher, Professor of Health Neuroscience, and he leads me to a nearby café to talk.

'I must say it was one of the rudest shocks I had, that science isn't just a sort of relentless progression,' Fletcher says some time later, describing an epiphany in his early career. Because it turns out that many key ideas in modern cognitive science were had more than a century ago. Popular books and articles today regularly claim that 'only now' are brain scans beginning to uncover the truth about complex sociocultural phenomena such as creativity or humour.[1] The ideas of theorists from bygone ages are either whitewashed from history or dismissed as unscientific guesses. But a closer look invites us to rethink the value of the different ways in which we understand the mind. We've seen how Cognitive Behavioural Therapy was inspired in part by ancient Stoicism, and how Stoicism itself is undergoing a revival. That's just one example of how many of our cutting-edge theories of the mind are themselves rediscoveries.

In the last few chapters we've looked in turn at ideas that come back even though it's not clear that they're right; ideas that come back even though they are definitely wrong; and ideas that might well be wrong but are nonetheless useful. Now we'll consider the return of a cluster of ideas that seem to render such questions entirely irrelevant. It doesn't even matter whether they're wrong or right. That's just not where their value lies.

Placebo ideas

Lately I've been working a nice psychological trick on myself. It used to be the case that, when I'd finished reading a book that I had to review for the newspapers, the last thing I wanted to do was sit down and write the review there and then. I might write up the notes I'd scribbled in the margins, but then I'd usually leave the actual writing of the review for another day. Until one afternoon, having finished a book, I said to myself: 'Oh, I'll just jot down some thoughts about the book in a random order while they're still fresh in my mind. But I'm not writing the review, so there's no pressure.'

Two hours later, of course, I'd written the review.

It's cute that it worked once. What's surprising is that it continued to work. Each time I'd promise myself that I didn't actually have to write the review, and I'd end up writing the review. There is no mistrust born from experience: I believe what I tell myself every time. Apparently I can fool myself in this way indefinitely.

This idea – that I'm not going to write the review, even though plainly I am – seems to belong to a class of ideas that are helpful even though they're not necessarily true. Let's call them *placebo ideas*.

Another candidate for a placebo idea, for instance, might be what is known as the disease theory of alcoholism. The idea that excessive consumption of alcohol is actually an organic disease is bound up with the history of Alcoholics Anonymous, and in recent years medical researchers have begun questioning many of that organisation's tenets, including the idea that absolute abstention is the right course of action for everyone, and the very 'disease' model itself.[2] Of course, there is a spectrum of human behaviour around any particular issue, and their variety is presumably related to differences in genes, environment, and (as you'll remember) neuroepigenetics. And we often choose to label the propensity to act towards the end of some particular spectrum as a 'disease', even if we are not sure of the precise causation. Whatever the truth of this classification, though, the disease theory could still be positively useful to someone with alcoholism – it could for instance provide the kind of psychic comfort that derives from knowing one's problem has been studied and identified (in the way that many people with mental-health problems feel more positive upon receiving a specific diagnosis), and that there are therapeutic steps one can take (even if there are not exactly twelve of them). So the disease model of alcoholism can help a person with alcoholism even if it is not factually accurate: it is a placebo idea.

The extent to which it is right to challenge that kind of placebo is a moral question in itself. The neuroscientist Marc Lewis argues that any kind of addiction is not a disease but a behavioural problem. But he understands the appeal of the disease label to addicts. 'What really moves me,' he has said, 'is the addicts who get in touch and say, "Don't take this away from me. If you take away the disease label, then basically I won't be able to get better, if you don't let me understand myself as having a disease."' Because the disease label is what protects them against being blamed for their addiction by society. 'They feel that if it is a disease, they don't have to feel that burden or shame, because it's not their fault. It's hard to pull the rug out from under that without causing some upset.'[3]

Another example of a placebo idea might be the idea that homosexuality is encoded in the genome: that gay people are, as Lady Gaga puts it, 'born this way'. When this idea first gained popular currency in the 1980s and 1990s, it not only helped some gay people abandon any interior shame that attached to their orientation, but it was an excellent defence against continuing moralistic prejudice. To the kind of homophobe who thought that gay people actively chose what he considered perversion, the gay person could respond that this was simply part of her biology. And so the 'gay gene', too, is a powerful placebo idea.

Such is the complexity of the human psyche, indeed, that it turns out that some ideas might be powerful placebos even when they are definitely false. That, at least, is one implication of modern research on psychotic delusions. But to understand that, we'll first have to reintroduce ourselves to a great nineteenth-century German scientist.

The gift of sound and vision

The word 'polymath' is overused, but Hermann von Helmholtz – whom we met briefly earlier, complaining about the tsunami of speculative science being published in his day – certainly counts as one. He did important work in the study of human vision and in physics, and he also invented what is arguably the first musical synthesiser. That invention came about after von Helmholtz pioneered the analysis of musical sounds into their component harmonic frequencies. Through elaborately structured listening experiments, he figured out that what made the characteristic sound, or timbre, of an oboe different from that of a flute or a violin was mainly due to the different sets of harmonic tones that the instruments generated above the fundamental tone, which is the 'note' played (usually the only tone we consciously hear). Then, out of little more than wood and brass, von Helmholtz constructed a series of tone generators built on tuning forks made to vibrate repeatedly with electromagnets. By playing different sets of forks simultaneously to recreate the harmonics he had heard in real instruments, he managed to create eerily instrument-like sounds, and even various recognisable vowel-sounds from human speech. 'The nasality of the clarinet was given by using a series of uneven partials,' von Helmholtz reported with satisfaction in 1875, 'and the softer tones of the French horn by the full chorus of all the forks.'[4]

Von Helmholtz thereby built a mechanical synth and made possible the work of his compatriots Kraftwerk a century later. In fact he continued in the relevant treatise to build an elaborate theory of correct musical harmony on top of his scientific observations, though that aesthetic scheme didn't really catch on. But one crucial point of his musical researches chimes with what became his more celebrated arguments in the theory of human vision. When we hear the sound of a violin, he said, we *sense* all the different harmonic tones, because we have dedicated structures within the ear that respond to different frequencies. (This is absolutely true, though the anatomy of his time was slightly off on the details.) But it is only when the mind combines all those different harmonics in its internal processing that we recognise the sound and *perceive* it as the sound of a violin. Perception, therefore, must itself involve some process of unconscious reasoning.

And what was true of hearing was true of vision, as von Helmholtz argued in his great *Treatise on Physiological Optics*. What the human eye *senses* are merely gradations of colour and light. In order to understand all this data, the mind must be working behind the scenes, pulling the available information together and drawing sensory inferences or 'unconscious conclusions' from it, so that it seems to us as though we perceive a table or a pint of beer.[5] The way we see objects in our everyday world, then, is not as though we're looking through a transparent window. And it is not merely a straightforward, dumb process of nerve excitation. The world must somehow be rationally reconstructed in our heads from ambiguous hints out there. Logical reasoning, then, if only unconscious, is a fundamental part of perception.

This idea was very controversial in the late nineteenth century, when many investigators assumed human perception could be reduced to a merely mechanical process. Only in the last few decades of the twentieth century, indeed, was von Helmholtz seen to have been way ahead of his time, as the new discipline of cognitive psychology took his ideas and ran with them. These days, the Cambridge neuroscientist Paul Fletcher says, von Helmholtz is widely referenced in papers and lectures about cutting-edge models of how the mind works. And his insight also gives us a clue as to what might be happening when the modern mind seems to go wrong.

Guess what?

Fletcher began his career in medicine, and then went into psychiatry. 'The first patient I ever saw as a medical student on my psychiatry attachment,' he says, 'had a really severe psychotic illness.' The man thought people on the television were talking directly to him, and saw hidden messages everywhere; he ended up mutilating himself. 'I was completely fascinated by the idea that you could create this fabricated world with no evidence, and then act on it in that way,' Fletcher says. Since then, a lot of his research has been conducted with people who have hallucinations. Normally we think of hallucinations – seeing something or hearing something that isn't there – as examples of something going wrong with the mind. But what if the mind *always* works this way?

'I'm very intrigued by the idea that something like hearing a voice when there's nobody there, or seeing a vision when there's nothing there, could be a product of the way the brain or mind has to work anyway,' Fletcher explains. 'It's not necessarily some gross derangement, it's actually quite a creative process. And one of the things von Helmholtz said

– which I think is absolutely key to understanding this process – he said: "Such objects are always imagined as being present in the field of vision as would have to be there in order to create the same impression on the neural mechanism." And what he was saying is, when you see something, or when you perceive something, you are imagining what *would have* to be there in order for you to have that experience.'

In other words, what I am doing when I 'see a table' is taking in the sense data about patches of coloured light and then reasoning backwards, guessing (in the optimal, successful case) that *if* a table were in front of me, *then* it would produce the same sort of pattern of coloured light I'm seeing. And so I conclude that I perceive a table.

That idea is 'slightly counter-intuitive at first', Fletcher says, but it was taken up again by the American philosopher Charles Sanders Peirce, often cited as the father of semiotics. Peirce introduced a third kind of logical inference. We're familiar with deduction (reasoning about what must follow from certain premises or facts), and Baconian induction (reasoning about what is likely to happen in the future given what we know about the past). Peirce's third form of reasoning he called abduction. 'Basically abduction is reasoning from the data that you have, to what could have been the cause of those data,' Fletcher says. 'Which is sort of what Helmholtz was saying perception is: it's an act of abduction. You have some data there, and you try and get in contact with what could have been the cause, through that.' Abduction, it seems, is necessarily going to be a fuzzy business.

'It's logically very flawed,' Fletcher says, 'because you're essentially saying: "There's something black over there; ravens are black; therefore it's a raven", or something along those lines – you're reasoning backwards. And as Peirce said, this is just a guess.' But that's okay, because according to Peirce guessing is the only way to make any progress in human understanding at all. He wrote: 'The truth is that the whole fabric of our knowledge is one matted felt of pure hypothesis confirmed and refined by induction. Not the smallest advance can be made in knowledge beyond the stage of vacant staring, without making an abduction at every step.'[6]

If we combine von Helmholtz's insight about the necessity for subconscious reasoning in perception, with Peirce's identification of abduction as the logical form of that reasoning, then we get a clear picture of how potentially unreliable our normal experience is. 'There's a whole mess of ambiguous, noisy, uncertain data coming in,' as Fletcher puts it, 'and all you've really got to make sense of it is what you already know. So it might follow from that that this balance between what's coming in and what's already there might be shifted, as a way of explaining hallucination.' If you're having a hallucination, then, your mind is not working fantastically abnormally; it's just that it is leaning more heavily than usual on what you already know to try to project some meaning on to the ambiguous storm coming from the senses.

So maybe hallucinating isn't some terribly weird deformation of the mind after all. Maybe it's just an occupational hazard of being a functional human being. Indeed, people who are prone to hallucinations are actually better, according to Fletcher's experiments, at some tasks involving interpreting noisy images. 'I think it's an exaggeration of a very useful but potentially misleading physiological mechanism that we all have,' Fletcher says, 'and I think you can trace that back to von Helmholtz's ideas directly.' About now one is tempted to imagine the great German playing a triumphant melody on his tuning-fork synthesiser in the sky. And in the mean time we might want to retune our easy notions of what is normal and what is dysfunctional. We're all just making guesses in the dark.

The uncertainty principle

So the idea that, when we look out at the world, we directly perceive what is really there – that is not true. But we could hardly function without it. It's a very deep-seated placebo idea.

The same sort of reasoning, Paul Fletcher says, can even apply to thinking about psychotic delusions, such as the idea that the television is sending you messages. The basic problem is that your brain is pretty insulated from reality: 'It is actually quite remote from its world, encased in thick bone and receiving only the noisy, ambiguous, inconsistent signals of its few senses.' So, many neuroscientists think that the brain's strategy is 'to integrate the signals that it's receiving with respect to the signals that it has received in the past', and thus to *predict* what is really going on.

But there are risks to such a strategy. 'If we superimpose our own predictions on input,' Fletcher says, 'then there is a danger that we are *creating* much of our perception.' (And this is indeed what happens if you put someone in a sensory-deprivation chamber: they begin to hallucinate.) In order to avoid this happening all the time, we must be sensitive to a 'mismatch signal', which 'tells us when what is predicted and what is actually there are differing too widely for it to be down to expected noise, or variance, or unreliability in the signal'. This mismatch signal, or 'prediction error', is 'the message that we need to update our knowledge to accommodate some change in the world'.

It follows that the brain might have the most success by following a very simple principle: always seek to minimise prediction error. To do this, of course, you also need the capacity to make predictions about predictions. 'At the risk of sounding Rumsfeldian,' Fletcher says – wryly noting how, as we have seen, Donald Rumsfeld was widely (and unfairly) mocked for this idea – 'you need to acquire an intricate knowledge of what you can expect to know or fully predict, as well as what you can expect *not* to know.' In other words, in mental life as well as in business and science, you need a good handle on known unknowns.

Paul Fletcher thinks of the brain as 'an entity that strives to minimise uncertainty'. And we're all familiar with uncertainty in daily life. 'A lot of misery,' Fletcher thinks, 'arises from striving to predict the unpredictable – and yet, given that we don't know what is truly unpredictable, we are doomed to generate, modify, and discard hypotheses, or to stick to ones that are imperfect but reflect the best that we can do.' Being uncertain is uncomfortable, and we tend to try hardest to minimise uncertainty when we are most stressed. 'Indeed,' Fletcher says, 'I wonder whether stress and uncertainty are almost synonymous at times.' Perhaps, he suggests, a given belief is most useful to a person not if it is true, but if it helps to minimise uncertainty. Or in my terms: how powerful it is as a placebo idea.

Delusions, Fletcher says, are special beliefs that 'are formally defined as arising without good evidence and fixed even in the face of contradictory evidence'. Fletcher agrees with those psychologists who see delusions as 'a desperate attempt to find relief in the face of horrible confusion and uncertainty', but adds that this is also 'what normal beliefs are prone to'. Indeed, he adds, 'given that non-delusional beliefs can seem equally irrational, one might ask what one has to believe for it to be formally defined as a delusion. I'm sorry to say that it can boil down to believing something that a sufficient number of other people don't believe.' Indeed, many researchers now think in terms of a psychosis spectrum on which everyone has a place, rather than a binary distinction between psychotics and 'normal' people.

So delusions, as well as hallucinations, seem to be caused by the brain working just as it normally does – only more so. This evidently complicates our assumptions about what is normal and what is abnormal. And just as a delusion is a helpful placebo belief for someone suffering terrible stress and uncertainty, many of our own apparently normal beliefs could be working in the same way. Beliefs can be placebos. But is belief in the power of a placebo itself a placebo? And might we have to reconsider our notions of how placebos work, how to use them, and how many we actually live by?

Please please me

The quantum physicist Niels Bohr had a horseshoe nailed over the front door of his country cottage for good luck. Seeing it one day, a surprised visitor asked: 'Surely you don't believe that nonsense?'

Bohr replied: 'They say it works even if you don't believe in it.'[7]

Bohr's joke was both clever and accurate. Indeed, one of the very odd things about the placebo effect is that a dummy pill has a measurable benefit even if you know it's a dummy pill. (Which is presumably why I can continue to trick myself into writing.) We are used to seeing people denounce various 'alternative' therapies as being in their effect 'no better than placebo', but this is slightly unfair to placebo. The placebo effect is mysteriously strong – and, even more mysteriously, it appears to be getting stronger.[8] If an intervention or drug is no better than placebo, that's already pretty good.

According to a widely cited but controversial meta-analysis, the placebo effect barely exists.[9] At the same time, many experiments have produced extraordinary results. For people in pain following oral surgery, the open administration of nothing more than salt water (saline), with the suggestion that it would reduce pain, was as effective as the covert administration (without telling them) of 6 to 8 mg of morphine. To be more effective than the placebo, the morphine dose had to be doubled to 12 mg.[10] So it seems that a placebo can be fully half as good as the most powerful painkiller known to man. For a while it was thought that this was simply due to the body producing its own painkillers (endorphins), but clever experiments suggest that the placebo response activates different internal chemicals depending on what drug the subject has previously been given. As the science writer Michael Brooks puts it, 'What everyone thinks of as "the placebo effect" turns out to be a whole array of different effects, each with a unique biochemical mechanism.'[11]

Other experiments infer that unexpected placebo effects are going on when people take well-established medications. Cunningly, researchers split anxious post-operative subjects into two groups. The first group was openly given diazepam (Valium) and told that they were receiving an anxiety-alleviating drug. They subsequently felt less anxious, as expected. The second group was covertly given exactly the same dose of diazepam (through a drip), but not told about it. These patients remained just as anxious as before. As one of the researchers, Fabrizio Benedetti, concluded, 'anxiety reduction after the open diazepam was a placebo effect'.[12] Yet if you tell people you are giving them diazepam but instead give them something totally inert, the effect won't be as large as if they'd had real diazepam – which, remember, doesn't work at all if people don't know they are getting it.[13] So somehow the placebo effect here depends on the patients' expectations *plus* a certain chemical.

Placebo is Latin for 'I will please', and until the middle of the twentieth century, when the 'placebo effect' was first named, the word placebo was used simply to denote a kind of quack remedy. It is defined, for example, in Robert Hooper's *New Medical Dictionary* of 1811 as 'an epithet given to any medicine adapted more to please than benefit the patient'. Now, of course, we know that the pleasing has concrete benefits. But some people already suspected as much, a very long time ago.

In Plato's dialogue *Charmides*, we learn of the principle used by the doctors of the Thracian king Zamolxis: the soul must be treated first if the head and body are to be well. And the way to treat the soul is by means of 'fair words', spells, or incantations; after which it is easy to coax the body back to health.[14] The 'fair words' of these physicians are presumably those that induce nothing other than a placebo effect in the patient.

The placebo effect has an equally powerful opposite, the *nocebo* effect (Latin for 'I will harm'). If they believe or expect something to go wrong, subjects in drug trials who are given placebos (sugar pills) can experience worsening symptoms; and powerful anaesthetics can lose their pain-relieving effect. This too has been intuited for a very long time. If curses can work, for example, it is probably because of a nocebo action.

As far as anyone can tell, then, placebo and nocebo responses are not caused by the agents (sham medicine, radiation) that apparently set them off but by a combination of causative factors that include the individual's own beliefs and a wide social context. They are, to this extent, psychosomatic. Which doesn't mean the experience, whether positive or negative, is any less real. Perhaps there should be nothing mysterious about such effects. Remember our neo-Lamarckian scientist, Isabelle Mansuy, pointing out that talk therapy might target the right brain areas more accurately than current medications. The reassurance and suggestion of a doctor, or a familiar pill, could also be targeting brain areas and altering them in helpful ways. Fabrizio Benedetti himself argues in this way: 'As the placebo effect is basically a psychosocial context effect, these data indicate that different social stimuli, such as words and rituals of the therapeutic act, may change the chemistry and circuitry of the patient's brain.'[15] If that is true of the placebo effect in medicine, it must also be true of placebo ideas. If an idea rearranges your mind, it must also 'change the chemistry and circuitry' of your brain.

Something very similar to the placebo effect is also put to good use in other areas of life – for instance, in restaurants. It is well established that people judge dishes with fancier names to be more delicious than exactly the same food described more simply, and the atmosphere of a restaurant also affects their perception of how good the food is. You can look at this cynically, as restaurateurs deliberately pulling the wool over diners' eyes for profit, but we should remember at the same time that eaters who are subconsciously manipulated in this way really *are* enjoying their food more than they otherwise would. Just as when medical suggestion can be as powerful as 6 mg of morphine, in fine dining too the placebo effect isn't a mere illusion.

In the same way, perhaps people who buy the expensive, colourful headphones of a certain brand really are having a richer listening experience, even though a frequency analysis might show that the luxury headphones reproduce sound no better than much cheaper models. The ideas that the food in a nice-looking restaurant is better, and that expensive headphones have better sound, seem to be perfectly harmless delusions. So if the placebo effect generally makes our lives better in all sorts of ways, why complain?

Well, at least in medicine, placebos do pose a lot of ethical conundrums. If, for example, you want to conduct a clinical trial in which a new drug for a life-threatening condition is tested against a placebo, then 'success' will mean that more people in the placebo group must die. For that reason, the standard these days is for trials of new treatments for such grave conditions to be conducted against the best current treatment, not against an inert substance (even though the inert substance might also have done good because of the placebo effect).

There are also ethical questions about how far doctors can go in wilfully harnessing the placebo effect. Would it be acceptable for a medical practitioner to knowingly prescribe a placebo? People tend to say no, since as a Parliamentary Commission in the UK pointed out, prescribing a placebo 'usually relies on some degree of patient deception'. But it evidently happens quite a lot anyway. In surveys of Danish and Israeli doctors, for example, more than half did it regularly.[16] Even though in some cases it can have negative effects for everyone. Consider the doctor who agrees to prescribe antibiotics for cold or flu – they won't do anything biochemically, because cold and flu are caused by viruses, and antibiotics kill bacteria. But patients often demand them, and they probably work because of the placebo effect – that is, when the infection doesn't run its natural course and clear up anyway. But increased use of antibiotics leads to increased antibiotic resistance among bacteria, and so to superbugs that are very difficult to treat.

It turns out, though, that telling someone you are giving them a placebo, and explaining how powerful placebos can be, can also induce a placebo effect.[17] In that case no one is being duped. And isn't it unethical to refuse to use a treatment that can be effective, even if we don't fully understand how it works?

The doctors of the Thracian king Zamolxis, who insisted that health travels from the soul to the body, found a common spirit in the French pharmacist Émile Coué, who as we have seen employed the principles of Stoicism in his own self-help techniques. Coué was also very interested in the placebo effect – or, as he put it, the power of suggestion. His pharmacist's shop was in the town of Troyes, and he observed that when he added little notes praising the medicine's efficacy for his patients to read, their condition seemed to improve more readily. So he began to do this systematically. (He also thought doctors should prescribe medicines to all patients even if they weren't necessary: that way a purely psychosomatic illness could be cured. It was probably lucky he didn't have a stock of antibiotics.)

Coué later explained his placebo theory in *Self-Mastery Through Conscious Autosuggestion*. He warned that the power of negative suggestion could have dire effects: 'If a doctor who by his title alone has a suggestive influence on his patient, tells him that he can do nothing for him, and that his illness is incurable, he provokes in the mind of the latter an autosuggestion which may have the most disastrous consequences.'[18] (Remember that an 'autosuggestion', on Coué's theory, is an internal unconscious idea that has powerful effects throughout the body.) Modern science agrees with Coué here: it is well known that a gloomy prognosis can exert a powerful nocebo effect.

If, on the other hand, Coué writes, the doctor tells the same patient 'that his illness is a serious one, it is true, but that with care, time, and patience, he can be cured', then the doctor 'sometimes and even often obtains results which will surprise him'.[19] Coué is being properly careful here with his 'sometimes and even often' – an optimistic chat with one's doctor is not likely to cure cancer. On the other hand, being deliberately misleading (if the doctor is sure the prognosis is hopeless) will not do the patient any favours either, for this patient would probably do better to get her affairs in order rather than assume she will be well again.

Coué was, even so, ahead of his time. He said that he would like to see the 'theoretical and practical study of suggestion' on the syllabus of medical schools. And modern researchers are indeed investigating the biochemical basis of suggestion – but they warn that such knowledge might be dangerous. 'Placebo research underscores the instability (or meta-stability) of the human mind and its somewhat dangerous tendency to be manipulated, particularly by verbal suggestion,' write Luana Colloca and Fabrizio Benedetti in *Nature*. 'For example, the assertion that placebos, fake therapies, fresh water and sugar pills could positively affect the brain biochemistry in the appropriate psychosocial context might lead to a dangerous justification for deception, lying and quackery.' These are examples of 'potentially negative outcomes' of placebo research. Therefore, 'If future research leads to a full understanding of the mechanisms of suggestibility of the human mind, an ethical debate will then be required, aimed at avoiding the misuse of placebos and nocebos.'[20]

Such a debate will become necessary after revising the conventional view that placebos *sort of* work by a kind of funny psychological trick, but that they're always inferior to real medicine. Placebos *really* work.

Practically speaking

The placebo effect acts not only with pills but also with talking cures. After all, what do both Cognitive Behavioural Therapy and Stoicism try to do? They try to help the subject eliminate harmful ideas – negative automatic thought processes; the 'bad nails' Coué talked about. Those ideas (whether they are true are not) are noxious – they are nocebo ideas. So they should be replaced by better, more positive ideas – whether or not those ideas in turn are true. In other words, these therapies want to install a set of placebo ideas in the patient's mind.

Some CBT sceptics, who favour the Freudian tradition, suggest that CBT only works because of the 'placebo effect'.[21] But how else could any talking therapy work? If you define the placebo effect as the use of the power of suggestion to produce a desirable change in the subject's experience, it's clear that CBT and psychoanalysis, when successful, must both be exploiting it somehow. It may be, in fact, that what is really powerful in any such therapy is the reassuring presence of someone who is paying sustained, friendly attention to you. That would explain what is suggested by some studies to be a 'Dodo bird effect', according to which all psychotherapies work as effectively as one another, regardless of their underlying theories. If that is true, then psychotherapy, of whatever kind, is a placebo. But again, how could it not be? The idea only sounds dismissive if we have not fully understood the extraordinary power of the placebo effect itself.

For a placebo idea, weighing reasons to chase down its truth or otherwise is beside the point. My own placebo idea to trick myself into writing is not predictably false – occasionally I have just jotted down some random thoughts and not actually written the entire article, and that is fine because it's all I ever promised myself I would do. What matters is that it often does help, and there is no downside to telling myself the story. And in general this is what should govern our attitude to placebo ideas: we judge them not by whether they are (as best we can determine) true or false, but by how useful they are to us. We treat the idea as a tool that must serve some useful function, otherwise it is to be discarded. Whether it actually reflects reality or not is irrelevant. Note that this is a much more radical position than David Lewis's rationale for accepting the existence of many worlds and infinite numbers of donkeys. Lewis says that the fact that this idea is useful is 'a reason' to think it is true. The person who embraces placebo ideas says that the fact that an idea is useful is a reason *not to care* whether it is true.

If this sounds dangerously postmodern, it's actually rather older than that. It is what Friedrich Nietzsche means in the epigraph to this chapter when he says: 'The falseness of an opinion is not for us any objection to it.' Nietzsche goes on, as he must:

> The question is, how far an opinion is life-furthering, life-preserving, species-preserving, perhaps species-rearing, and we are fundamentally inclined to maintain that the falsest opinions [. . .] are the most indispensable to us, that without a recognition of logical fictions, without a comparison of reality with the purely *imagined* world of the absolute and immutable, without a constant counterfeiting of the world by means of numbers, man could not live – that the renunciation of false opinions would be a renunciation of life, a negation of life. *To recognise untruth as a condition of life*; that is certainly to impugn the traditional ideas of value in a dangerous manner, and a philosophy which ventures to do so, has thereby alone placed itself beyond good and evil.[22]

Well, if placebo ideas are indispensable, as I have been arguing, then perhaps untruth really *is* a condition of life.

A similar view was held by the psychologist William James. He supposed that a reasonable way to decide whether to follow the precepts of some religion, for example, was not to wonder whether the religion was true, but to ask whether following those precepts would make one's own life better. That is what became known as 'pragmatism' in philosophy. 'Pragmatism asks its usual question,' James explained. '"Grant an idea or belief to be true," it says, "what concrete difference will its being true make in any one's actual life?"'[23] If it makes a positive concrete difference, by all means adopt it. It seems clear, for example, that those who find comfort in religious faith at times of crisis are benefiting from the very real help of a placebo idea. It may follow that certain religious sub-traditions, on the other hand, are not placebos but definitively nocebos, at least on a social level. (The idea that it is virtuous to blow oneself up and thereby kill whoever else is nearby does, after all, seem to be actively harmful: if not to the holder, who is convinced of thereby attaining paradise, then certainly to others.)

Some anthropologists divide certain social behaviours observed in tribal societies into 'placebo rituals' (for instance, faith healing) and 'nocebo rituals' (curses). It would be hubristic to imagine that we industrial moderns don't do very similar things. Indeed, there are doubtless ideas that function for us as social placebos. Take the idea that it is important for you to vote in a general election. Well, in one sense it isn't – because it definitely won't make any difference to the outcome. No major election, of the direct presidential kind (as in the US) or the local-representative kind (as in the UK), has ever been decided by a margin of only two votes. So your vote is literally pointless, and casting it is a placebo ritual. However, if everyone acted on this true belief and declined to vote, then democracy would fall apart. Some states get round this problem by making it mandatory to vote, as in Australia. Others use a softer strategy, simply by encouraging the placebo idea that your vote matters. And so politics can trundle on. In these ways, rethinking can remind us that what is most important in social and cultural ideas, too, is often not their truth or falsity, but the effect they have on people. The placebo effect works on groups as well as on individuals.

As if!

Pragmatism, as William James formulated it, also feeds into the CBT system. The most popular modern handbook recommends that if you habitually get angry in certain situations, for example, you should ask: 'Is it to my advantage to get mad?'[24] Perhaps, in certain situations, it will be helpful. But in most it won't. So the anger, being disadvantageous, should simply be abandoned. Readers are invited to perform a 'Cost-Benefit Analysis' on all their negative thoughts in this way.[25] Pragmatism also informs Paul Fletcher's suggestion that we consider a belief's usefulness not according to its veracity, but according to its capacity to minimise uncertainty.

William James was right about something else too, though just how right wasn't appreciated until psychologists belatedly began to dust off and test his idea eighty years later. And what this strand of research seems to indicate is that going to all the trouble of changing the way you think, in the manner of a Stoic or neo-Stoic, might be unnecessarily difficult. Perhaps all you have to do is change the way you *act*.

You've probably heard of the idea that smiling makes you happy. But a quick grin won't do. Try it now. Relax your forehead and eyebrows. Stretch the corners of your mouth out towards your ears in the widest possible smile, making sure that the corners of your eyes crinkle too. And slightly raise your eyebrows. Now hold this position for a full count of twenty seconds.

Hello again! What's up? Well, most people report that, after doing this, they do actually feel happier, in a slightly silly way, for no good reason. And that is exactly what William James would predict. For he thought that our usual way of understanding emotions had things backwards. We think we run away from a bear because we are scared. James says we get scared once we find ourselves running away from a bear. The unconscious mind notes physiological changes in the body, interprets them, and then produces the appropriate emotion. According to more recent research, in fact, the very same physiological changes in the body can be interpreted by the mind to produce a variety of different emotions depending on context. So, as the psychologist Richard Wiseman explains, 'the same thumping heart can be seen as a sign of anger, happiness or love', depending on the interpretation we are predisposed to use at the time. Wiseman has written a very interesting neo-Jamesian self-help book based on the principle that if this is true, it follows that changing one's behaviour should be the most direct way to change one's feelings. He calls this the 'As If Principle', after James's suggestion: 'If you want a quality, act as if you already have it.' In other words, 'to calm down, act like a calm person'; to feel gregarious, mingle with people; to feel determined, make a fist. And so forth.[26] There is, indeed, a grand debate within modern CBT itself as to whether it is the C (the cognitive) or the B (the behavioural) aspect of the therapy that matters more. Jamesians argue the latter, convinced that it is behaviour that generates feeling.

William James's idea that emotion is essentially an unconscious inference from bodily states is still taught today on the Natural Sciences undergraduate course at Cambridge, says Paul Fletcher. 'There's an awful lot of good evidence that it's the case,' he says. But it's pretty hard to adopt as a way of living one's life. 'I quite like to talk to the students about this,' Fletcher says, 'because everybody accepts the logic of it but nobody ever truly *believes* it.'

Indeed, James's idea is still a shocking and outrageous claim more than a century later. We are so used to the idea that our emotions are our inner reality, our guarantor of personal integrity. If nothing else can be trusted, our feelings can. Perhaps more than ever, in an age of identity politics and trigger warnings, our emotions seem to represent the peak of what can count as real and authentic. So it's extremely difficult to make the gestalt leap to identifying our emotions as mere reactions to physiological processes. Or to follow the advice of the great Leonard Cohen: 'I don't trust my inner feelings, inner feelings come and go.'[27]

William James himself chose never to conduct experiments to confirm his theory of the emotions because he, er, hated experiments. 'The thought of brass-instrument and algebraic-formula psychology fills me with horror,' he wrote charmingly.[28] (Score another point for armchair theorising or rationalism.) In modern times, though, many researchers have tested James's ideas. And they do seem to hold up.

One particularly fascinating experiment that sought to extend them further was reported in 1962 by the psychologists Stanley Schachter and Jerome E. Singer.[29] In dreaming up their experiment, Paul Fletcher explains, 'They reasoned that it's perfectly valid to assume that physical sensations produce emotions, but that there's a cognitive overlay as well that modulates that relationship.' How to test this idea? Well, Schachter and Singer gave one set of volunteers a shot of adrenaline, which gets the heart beating faster and the palms sweaty. These physical symptoms, Fletcher says, 'are quite ambiguous – they could be exhilaration, they could be fear, they could be anger'. And then the experimenters told each group of subjects different things. One group was told that they'd had a drug which would make their palms sweat and their heart race. Another group was given false information – 'They said "It will make your toes go numb" or something.' A third group was given adrenaline but not told anything about it, while a fourth (control) group was given a placebo which didn't induce any symptoms (because they weren't told to expect any). Then each group was put in a room together with an actor who was behaving either very happily or very angrily.

The question was this: would emotional contagion take place more easily in people who didn't know *why* their hearts were racing? Would they 'catch' the actor's delight or anger, and become euphoric or furious themselves, in an unconscious attempt to explain their physical symptoms? Yes: that is exactly what happened. Meanwhile, the people who *knew* that they'd been given adrenaline did not so easily catch the other person's mood. 'If you've got these physical sensations, but on top of that you have a cognitive model for why that's happening,' Fletcher says, 'you're relatively resistant to the emotion.'

This is a classic experiment that seems to confirm what became known as the 'two-factor theory' of emotion. (The two factors are physiological arousal and cognition.) 'It is interesting,' Fletcher says thoughtfully of the experiment. 'The only problem is, everyone I've ever spoken to says it's never been replicated. It's always taught, but never replicated.

'And the reason it's taught is because it's so beautiful.' Fletcher smiles. 'We're not immune to a good story in science, are we?'

We know that the mind influences the body; James's bet is that the body also influences the mind. A similar notion informs the wildly successful TED talk by the Harvard social psychologist Amy Cuddy, entitled 'Your Body Language Shapes Who You Are', which has had four million views. She describes how modern research suggests that adopting a 'power pose', an expansive and strong physical posture, for just a few minutes makes us feel more confident, and can help us be more successful. This is explained, in modern scientific terms, through hormonal changes in the brain.[30] But the concept that the body's stance affects the mind is very old news in other traditions – especially the ones that never separated mind from body in the first place. A powerful physical pose has beneficial psychological effects? Someone who practises the Chinese martial arts will not be at all surprised.

It could be that the neo-Jamesian two-factor theory of emotions itself is just a placebo idea. But if it's useful, there's nothing wrong with that. The pragmatist William James himself would surely approve. His theory encourages us to rethink our major cultural assumption about the primacy of feelings, and relegate them to the status of mental effects that arise from what we actually do. And, funnily enough, that can end up making us feel better. So why not do it?

Free won't

Another powerful placebo idea that we can't help but live by is the subject of an ancient controversy that has also in recent years become, somewhat surprisingly, the subject of renewed public debate. It is the question of free will. Among many others, the writer Sam Harris has issued, in *Free Will*, an elegant denunciation of the idea, and the philosopher Julian Baggini, in *Freedom Regained*, has promised to recuperate it. The problem is that free will, in the way most people understand it, seems to be incompatible with the findings of modern science. So Harris, for example, argues that we should just give up the notion. We're not free in the way we suppose, he argues, and it doesn't matter that we aren't. But the question remains: do we *know* that we *don't* have free will?

Here I am using 'free will' in what philosophers dismissively call the 'folk' sense, or the 'libertarian' sense. I have free will in this sense if, at a given point in the past, I could have done otherwise than I actually did – even assuming that everything else about the universe had been the same. In other words, at any given moment, taking into account all the micro-physical facts about the world and my body and my brain, I still have a choice as to what to do next. The problem is: where does this choice come from? It seems that states of the universe (including states of my body and my brain) are inexorably caused or *determined* by previous states of the universe (including states of my body and my brain). There is an unbroken chain of causation from one moment to the next. What within me is able to jump in and disrupt the chain, sending it off in a different direction?

The answer, of course, used to be the soul. But apart from that, there seems no other candidate for a magical choice-making part of ourselves that is not subject to demonstrable causation, or determinism. We cannot escape the prison of the laws of nature. (Even if, as Rupert Sheldrake and others suggest, they are just well-ingrained habits.) So we simply can't have free will. This idea goes back at least to Democritus: since everything was atoms and void, he thought that free will was just an illusion. And the modern notion of quantum indeterminacy – that, deep down, reality is not strictly determined but probabilistic or aleatory, subject to chance – does not help us here. Mere randomness down among the particles can't give us freedom to choose. To someone mired in existential gloom it will not help to say that this is just an aleatory crisis rather than a deterministic one.[31]

There is, moreover, a series of famous experiments in psychology that are often said to prove the non-existence of free will. In the 1980s, Benjamin Libet and his team of researchers asked volunteers to spontaneously perform a physical action – lifting a finger or flexing a wrist – and note the exact time at which they decided to do it. At the same time, Libet was monitoring their brain activity. He found that a 'readiness potential' in the neurons preceded the conscious decision to move by around a third of a second. Many subsequent discussions of these and similar experiments draw the baleful conclusion that what the subjects experienced as a free choice was in fact already determined by unconscious processes in the brain – so there can't be any such thing as free will after all. Yet Libet himself did not interpret his own experiments in quite this way: he suggested that a person retained the power of veto over actions that the unconscious brain had initiated, and so was still free. (Some people call this idea not 'free will' but 'free won't'.) Other observers consider all such experiments inconclusive because of methodological issues, such as how difficult it is for a person to report the precise timing – to within a few tenths of a second – of when she makes a decision. So for many philosophers and neuroscientists, Libet-style experiments have not resolved the issue.[32]

Whatever the interpretation of such studies, however, the general argument from determinism – that we are physical organisms in a law-governed universe – just seems to many people like a knock-down argument that free will cannot exist. And yet – remember how challenging William James's theory of emotion was: you can accept the logic of the argument, but it's really difficult to *accept* that your emotions are not a meaningful part of your true self. The case is similar with the problem of free will: one may accept the logic of the above argument, and yet it's very hard – maybe outright impossible – to live your life as though you *don't* have free will. As the philosopher John Searle puts it: 'If, for example, I am in a restaurant and I am confronted with a menu and the waiter asks me what I would like, I cannot say "I'm a determinist, I'll just wait and see what happens", because even that utterance is only intelligible to me as an exercise of my free will.'[33]

A broad current in philosophy, called 'compatibilism', attempts to solve the problem by adopting a much more limited idea of free will than our ordinary one. Crudely speaking, it says that your actions are free as long as they are not externally coerced and they are in accordance with your character and desires. To some people, this looks like a cop-out: after all, what formed your character and desires? On the deterministic view, just a preceding chain of strict causes over which you had no control. But the compatibilist just shrugs and says sorry, that's all you're going to get. Daniel Dennett is a compatibilist. He acknowledges that 'compatibilism may seem incredible on its face, or desperately contrived, some kind of a trick with words', but points out that it is the view held by 'the majority of philosophers'. (The survey he cites says that compatibilism is the view of 59 per cent of professional philosophers, which you may not find an overwhelmingly impressive majority.) To add insult to statistics, Dennett impishly adds that when 'everyday folk' – or, the vast majority of people who don't work in university philosophy departments – talk about free will in the large sense I have described it, they 'mean something demonstrably preposterous'.[34]

Compatibilism, however, doesn't conserve what is important to many benighted 'everyday folk' in the idea of free will – that we could have done differently. So whether we have free will or not, the *idea* that we have free will seems to be a very important placebo. It might be a social placebo as well: some research suggests that people who are persuaded to believe they don't have free will are more likely to act dishonestly.[35] (If everyone ceases to believe in free will, of course, our systems of justice will require a thorough overhaul: if a criminal is just the slave of impersonal physical forces, the very notion of moral responsibility looks shaky, and punishment seems simply sadistic.) But there are upsides to denying free will too, at least for the individual so persuaded if not for everyone else. If you can become convinced that at no point in your past could you ever possibly have acted differently than you did, then you will experience no regret or shame. (What would be the point? At any given moment it was literally impossible for you to do otherwise.) Some people, such as the philosopher Ted Honderich, do claim to have convinced themselves of this view, and it is tempting to envy them. But most of us will continue to think we are free in the normal sense.

Consider, though, an alternative: perhaps 'real' free will is a black box. It is rejected by most scientists and philosophers today because no one can see how it is compatible with our understanding of how the universe operates. But that, of course, does not necessarily make it wrong, any more than an inability to understand how Lamarckian inheritance could work made it wrong for a century until epigenetics came along. The truth is that we still don't know what choice really means on the neurochemical level. And what Daniel Dennett did not say was that, according to the same survey of modern philosophers he cited, nearly 14 per cent of them hold the view known as libertarianism – which is not the same as the political idea of the same name, but means they believe in real free will. (Meanwhile, 12 per cent say there is no free will at all, and 15 per cent plump for the intriguing category 'other'.)[36] People currently can't see how free will could work. But we should only be confident in rejecting it if we are very confident in our ideas about the nature of fundamental reality. And that's a harder sell – especially to people whose business it is to investigate fundamental reality. Physicists are more inclined to give weird ideas a hearing than the insistent biologists of our age, perhaps because physicists are more intimately acquainted with just how weird reality seems to be.

The quantum physicist Erwin Schrödinger was one of them. In his classic series of lectures entitled *What Is Life?*, he decided that 'quantum indeterminacy plays no biologically relevant role' in the operations of the mind. Indeed, he was sure that 'my body functions as a pure mechanism according to the Laws of Nature'. And yet he was *also* sure that he had free will: 'I know, by incontrovertible direct experience, that I am directing its motions, of which I foresee the effects, that may be fateful and all-important, in which case I feel and take full responsibility for them.' The only way, as he saw it, to reconcile these two facts was to say: 'I – I in the widest meaning of the word, that is to say, every conscious mind that has ever said or felt "I" – am the person, if any, who controls the "motion of the atoms" according to the Laws of Nature.'[37] A grand thought – and one that, as Schrödinger pointed out, dates back two and a half millennia to the Upanishads. Mystical, yes; but arguably no more mystical than determinism itself. (Consider: if we adhere to strict determinism, we are compelled to believe that Mozart's Symphony No. 40 somehow already existed, in potential but precise form, at the moment of the Big Bang – because it was inevitable and, in principle, predictable, that it would be written.)

Perhaps there are still other possibilities. It could be that what seems like the exhaustive dichotomy of determinism versus probabilism is missing some third alternative that no one has thought of yet. Or perhaps we ought to rethink a thought of that sclerotic and abusively hilarious nineteenth-century German philosopher, Arthur Schopenhauer. He proposed that a *will* to continue existing was what drove everything in the universe. (Nietzsche's later development of this idea was the 'will to power'.) Well, let's take this seriously, and then make the same kind of argument that Galen Strawson offers for panpsychism.

Let's say, then, that we *know* we have free will, but we have no idea how it could be generated from the interactions of matter. We also reject 'spooky emergence', the hand-waving idea that something utterly unlike matter – whether mind or will – can just 'emerge' from matter in some way we can't even begin to describe. It follows that will, just like mind, must be a fundamental property of *all* matter, deep down. It might be that panpsychism itself implies this, since it is difficult to conceive of consciousness without will. (Though the lassitude of a procrastinating writer might be a strong counterexample.)

In his autobiography the theoretical physicist Freeman Dyson explicitly connected his affirmation of panpsychism with the problem of free will. 'I cannot help but think that the awareness of our brains has something to do with the process that we call "observation" in atomic physics,' Dyson wrote. 'That is to say, I think our consciousness is not just a passive epiphenomenon carried along by the chemical events in our brains, but is an active agent forcing the molecular complexes to make choices between one quantum state and another. In other words, mind is already inherent in every electron, and the processes of human consciousness differ only in degree but not in kind from the processes of choice between quantum states which we call "chance" when they are made by electrons.'[38]

Very well, then. If mind is already inherent in every electron, so is free will, in order to enable the electron to make those 'choices' between quantum states. If you can have panpsychism, you can have a neo-Schopenhauerian *panwillism*. I don't know if that's true, but it makes me feel better.

The truth of whether we have free will or not, that old controversy given new life in the age of modern neuroscience, arguably doesn't really matter. Everyone – apart from a few miraculously hard-headed souls – is going to carry on living as though they do. Our very freedom may just be a placebo idea, but to recognise that truth can itself be a liberation for some. It is also difficult to abandon the placebo idea that we perceive the world directly, even though the return and upgrading of von Helmholtz's theory of perception implies that we don't: that all of us are just making guesses and predictions, for better or worse. And what about William James's theory of emotions as secondary effects, rather than primary sources of internal meaning? The wholesale embracing of this old idea, in particular, is likely to meet stiff resistance in an age obsessed with the individual's authenticity and personal growth. Yet it too could be a very powerful placebo for those adventurous souls willing to think it over again. In such cases the motivation for recovering past ideas is not necessarily 'Is it true?' but something more visceral and personal: 'Can it help?'

Part III: Prognosis

11

Utopia Redux

Which old ideas could we resurrect to improve our world right now?

'The power of vested interests is vastly exaggerated compared with the gradual encroachment of ideas.' – J. M. Keynes

Forgive me if I had been expecting bad hair and ill-fitting tweeds, but it turns out that the opening party for the annual Historical Materialism conference in London is held in a hipster warehouse space crowded with elegant young intellectuals. I get into beer and conversation with a more senior chap at the bar who knew the great Herbert Marcuse, the German-American philosopher and public intellectual who found fame with countercultural classics such as his 1964 book *One-Dimensional Man*. (He is often said to have founded or spearheaded the 'New Left' movement.) My new friend explains how delighted he is that his students today recognise so much of their own concerns in Marcuse's thrilling attacks on the psychological and social cost of consumerism, the unfreedom of modern 'administered life', and the way technology can operate 'as a form of social domination'.[1] And that was half a century ago. You can imagine what Marcuse might say about mob shaming on social media, and the pervasive state surveillance of our communications.

Yes, Marcuse is back, and so is the critique of our dominant economic model in general, to judge by the packed lecture rooms at the Historical Materialism conference the next day. Indeed, since the global financial crisis of 2008, followed by the Occupy movement and the surprise bestsellers by David Graeber about debt and Thomas Piketty about inequality, people all over the world have been putting the politics back into political economy. In Greece, with the election of Syriza and the short reign of dissident economist Yanis Varoufakis as its finance minister; in Britain, with the surprise election in 2015 of Jeremy Corbyn as leader of the Labour Party. In 2009, the billionaire George Soros had funded an Institute for New Economic Thinking. One of its contributors was Adair Turner, formerly the head of Britain's financial regulator, who announced: 'We know there needs to be a complete rethink of how we talk about economics and train economists.'[2] The Bank of England even started to hire non-economists. Its incoming director, Mark Carney, explained: 'We know we need to get more diversity of thought.'[3]

The very notion that a financial crisis – or even just a run on a bank – could be possible in a modern economy was itself the spectacular return of an idea long thought dead.[4] And since the crash, more and more people have been saying things like this: 'When the capital development of a country becomes a by-product of the activities of a casino, the job is likely to be ill-done.' Or this: 'The outstanding faults of the economic society in which we live are its failure to provide for full employment and its arbitrary and inequitable distribution of wealth and incomes.'

Those quotations are characteristic of the sort of criticism of 'casino banking' and inequality that became newly popular after the global financial crisis of 2007. But they were written eighty years ago, by John Maynard Keynes.[5] It took him a while to come back. Partly because, as the Nobel laureate in economics Paul Krugman writes, there was in the US a concerted effort to suppress Keynesian thought in the teaching of economics after the Second World War.[6] Why would some people want to suppress it? Krugman explains: 'Keynesian economics, if true, would mean that governments don't have to be deeply concerned about business confidence, and don't have to respond to recessions by slashing social programs. Therefore it must not be true, and must be opposed.'[7] In other words, '[B]usiness interests hate Keynesian economics because they fear that it might work – and in so doing mean that politicians would no longer have to abase themselves before businessmen in the name of preserving confidence.'[8]

Meanwhile, Marxism itself is back again too – as usual. Jacques Derrida wrote in 1993: 'After the end of history, the spirit comes by *coming back*, it figures *both* a dead man who comes back and a ghost whose expected return repeats itself, again and again.'[9] Every new financial crisis, indeed, shows that many of the dominant assumptions of economics are just false (as we have seen, they are zombies), and the neo-Marxian alternative, no matter how often people claim that it is refuted by history or reason, is always waiting in the wings.

In 1798, Thomas Malthus warned against unreasoning optimism about the perfectibility of human society. 'A writer may tell me that he thinks man will ultimately become an ostrich,' he wrote. 'I cannot properly contradict him. But before he can expect to bring any reasonable person over to his opinion, he ought to shew that the necks of mankind have been gradually elongating, that the lips have grown harder and more prominent, that the legs and feet are daily altering their shape, and that the hair is beginning to change into stubs of feathers. And till the probability of so wonderful a conversion can be shewn, it is surely lost time and lost eloquence to expatiate on the happiness of man in such a state; to describe his powers, both of running and flying, to paint him in a condition where all narrow luxuries would be contemned, where he would be employed only in collecting the necessaries of life, and where, consequently, each man's share of labour would be light, and his portion of leisure ample.'[10]

Malthus has a fair point, beautifully expressed. But there are ideas about reworking our political economy – modern versions of very old ideas – that do not depend on avian metamorphosis. They are very simple. And they invite us to rethink what should and shouldn't be dismissed as too simple to work in reality.

Back to basics

One thing Herbert Marcuse wrote that resonates with a lot of people today is this, from *One-Dimensional Man*: 'Economic freedom would mean freedom *from* the economy – from being controlled by economic forces and relationships; freedom from the daily struggle for existence, from earning a living.'[11]

How to achieve such freedom for everyone? One answer is simple: just give everyone enough money to live on, gratis. Well hang on, maybe that's *too* simple. Too radical, in both senses: threatening to established interests, and naïvely hoping to attack the root of the problem. It could never actually work. Right?

In fact, the idea of giving enough money to everyone to live on is another old idea, long thought ridiculous, that is now back on the political agenda in countries all over the world. It usually goes by the name of Universal Basic Income, and it does exactly what it appears to say: the idea is that the government gives enough money to live on to every single citizen, rich and poor. No more means-tested welfare, just a lump monthly sum to all, which can be paid for through increased corporation or other taxes. People can then choose how to spend their time – some will still want to trade stocks to become rich, others might prefer to become reiki therapists.

Barbara Jacobson is a droll Canadian who works as a benefits advisor in Fitzrovia, central London. She has been a housing and benefits activist in the city since 1982, and promotes UBI through the volunteer organisation Basic Income UK. On the piazza of the British Library in London, we are joined by her fellow volunteer Marlies Cunnen, who is originally from the Netherlands. Over lattes and strawberries, generously brought along by Jacobson, we discuss how this is another idea that has come back multiple times. Will it finally stick?

Start by imagining what it would be like to have a basic income. What would you do? Quit your job and retrain? Or just devote yourself to painting or yoga? One thing to this idea's advantage is that it tends to personalise otherwise abstract matters. 'The problem with talking about monetary or banking reform,' Jacobson points out, 'is that people's eyes glaze over almost immediately, and a real conversation about pros and cons is quite difficult. I find with basic income that often people have a visceral reaction to it, and whether pro or anti, it at least starts a conversation.'

The conversation about something like UBI dates back at least to Thomas Paine's 1796 pamphlet *Agrarian Justice*. In it, he addresses the revolutionary French government and points out that poverty 'is a thing created by that which is called civilized life. It exists not in the natural state.' To alleviate it he suggests that everyone should be given a capital endowment upon reaching the age of twenty-one, to reflect their 'equality of natural property' – the common inheritance of land and air – and a pension when they are fifty. 'It is not charity but a right, not bounty but justice, that I am pleading for,' Paine adds.[12] Similarly, the French philosopher Charles Fourier thought that civilised society violated everyone's natural right to hunt, fish, and graze animals on common land, so that in return the government owed to the poor citizens a guaranteed 'minimum' of subsistence.[13] (For this sort of thing Fourier was dismissed by Marx as an exemplar of 'utopian socialism', though many of his other ideas don't look so silly now. Fourier is credited, for example, with coining the term 'feminism' in 1837.) But it was probably the Belgian writer Joseph Charlier, from 1848 onwards, who proposed the first recognisably modern scheme of unconditional basic income for all.[14]

Only in the twentieth century, however, did a UBI begin to be taken seriously in politics. In fact, we are currently living through the third revival of the idea in a hundred years. Discussions of it went on during the 1920s, and again in the 1970s, but they were abandoned or at most relegated again to the fringes of academia. 'We have a supporter who is quite elderly and was involved with the Heath government,' Jacobson says, and around 1972 or 1973 the Heath government was thinking seriously about instituting a Basic Income. 'That was around the same time that Nixon was looking at a guaranteed minimum income, or a negative-income-tax proposal in the US.' But then the Heath government fell, and within a few years giving people money for nothing became a deeply unfashionable option. 'All of those forward-looking demands died with Thatcher,' Jacobson says, 'and everyone went on the defensive.'

And then, after the financial crisis, Basic Income came back again, in the most telegenic manner yet. In the autumn of 2013, demonstrators dumped eight million coins outside the Swiss parliament. The figure represented one coin for each citizen, and it was the opening salvo in a UBI movement spearheaded by Enno Schmidt, an artist. The Swiss activists say that a UBI will enable everyone to 'lead a life fit for human beings and to take part in public matters', and they have collected enough signatures to trigger an official referendum on whether the government will adopt UBI. It might not pass, but they've managed to kick-start a very interesting mainstream conversation.

Enno Schmidt has done a lot of interviews with the American press, and recalls that the reaction was often, in précis: 'Should every American live like this stupid painter from Switzerland?' Does he mind being called a stupid painter? 'No,' he told an interviewer. 'I'm proud of this because to do something good, you have to be a bit stupid in that moment. Being a bit stupid [allows] you to see more. Don't be too intelligent – because then you can think of every possible objection.'[15] This might be summed up as: 'Don't overthink; rethink!'

The idle poor

Silicon Valley is a world away from European socialist activism and conferences on Marxism, yet – perhaps surprisingly – UBI is also gaining ground in the rarefied world of technology venture capitalists. What, after all, will happen when robots destroy all the jobs? How will people live? This is the question that is driving tech UBI proponents in America, from all along the political spectrum. The idea has been taken up by the Netscape co-founder Marc Andreessen and Tesla Motors' principal software engineer Gerald Huff, among many others.[16] Sam Altman, president of the start-up incubator Y Combinator, has pledged to spend tens of millions of dollars on research into the possibility of Basic Income in the US. 'I'm fairly confident,' he announced on Y Combinator's website, 'that at some point in the future, as technology continues to eliminate traditional jobs and massive new wealth gets created, we're going to see some version of this at a national scale [. . .] 50 years from now, I think it will seem ridiculous that we used fear of not being able to eat as a way to motivate people. I also think that it's impossible to truly have equality of opportunity without some version of guaranteed income. And I think that, combined with innovation driving down the cost of having a great life, by doing something like this we could eventually make real progress towards eliminating poverty.' It is a stirring aim, even if some observers might suspect that Silicon Valley's motivation for embracing the idea is one of cynical self-exculpation: after all, these are the people designing the robots and software that will destroy all the jobs.

But maybe there's a deeper problem. If UBI has been repeatedly proposed and rejected over centuries, perhaps something is fundamentally wrong with the idea. What, I ask Jacobson and Cunnen, are the typical objections they meet from people when talking about it? 'Nobody will do anything any more!' Cunnen says. 'That's the usual one,' Jacobson agrees. 'But then,' Cunnen adds, 'if you ask them, "Is that what *you* would do, just watch TV?", they say, "No! But everybody else would."'

Ah, the eternal problem of others not being as respectable as oneself. Would *you* just sit around and watch TV? Or would you switch careers for your dream job? Or paint, and walk people's dogs?

Charles Fourier himself anticipated this very objection. Guaranteed a living, he suspected, the common people would want to work little if at all, so some kind of regime of attractive but mandatory work would have to be constructed. But what we know now, which Fourier didn't, is what actually happens when a UBI is tried: pilot studies have been conducted in many villages and towns in different countries, and it's just not the case that everyone becomes a dozy idler.

For five years in the 1970s, a UBI project called Mincome ran in the Canadian town of Dauphin, Manitoba. The results were analysed by the health economist Evelyn Forget, who found that only teenagers and mothers of young babies worked less than they had previously.[17] Further,

Jacobson explains, Forget 'found massive health benefits, in terms of people not needing to go to hospital, reduced domestic violence, reduced criminality'. And educational benefits, too: 'teenagers staying in high school longer, or actually finishing high school where they might not have'. So another argument in favour of UBI, she says, is that in Britain, 'It would actually save the NHS a lot of money.'

What's more, as Jacobson points out, 'Real life shows that you need money in order to work. You actually need the money first' – for travel, clothes, food, and so on. 'That's actually how [the German entrepreneur and UBI promoter] Götz Werner argues for it. You actually need to be paid and then you can go and work afterwards.' And the work people will do when supported by a UBI will potentially be far more interesting. 'I think it's a bad use of a human to spend twenty years of their life driving a truck back and forth across the United States,' the New York venture capitalist Albert Wenger told the writer Farhad Manjoo. 'That's not what we aspire to do as humans – it's a bad use of a human brain – and automation and basic income is a development that will free us to do lots of incredible things that are more aligned with what it means to be human.'[18] The fact that technology eliminates jobs is alarming – unless the people who used to do them have a safety net, and so can be considered as freed to do more interesting work, rather than thrown on the scrapheap.

If, then, we agree that a UBI will not make everyone indolent, that it will in fact help people to work, and might even enhance public health, another question arises. How much, exactly, should it be? In the 2015 UK general election, the Green Party proposed a UBI of £80 a week, carefully costed to be affordable under current tax arrangements. But no one can live on that, so it wasn't a particularly thrilling policy prospect. Discussions in the American tech sphere usually start at around $1,000 a month. But of course higher is always better, at least for the recipients. 'When we got together,' Jacobson says, 'it was really quite apparent that people were not gonna get out of bed for less than £12,000 a year, okay?' She laughs. So the Greens' proposal was uninspiring. 'In our literature,' Jacobson says, 'we've tried to target other sorts of taxes, whether land-value tax or taxes on corporate profits – other sources of money – and use that for basic income.'

So what is the right level of income for everybody? 'Um, personally . . .' Jacobson starts, before Cunnen jumps in: 'As high as possible!' Jacobson agrees. 'But I think a good place to start would be the Living Wage, so that's £14,500.'

Another problem with the Green Party's proposal, Jacobson says, was that it left the current benefits system in place. And getting rid of it is one of the advantages of a UBI. Indeed, there has long been a strong conservative case for a UBI – advanced by, among others, the economists F. A. Hayek and Milton Friedman in the twentieth century – which is that it eliminates the bureaucratic waste involved in administering unemployment benefits, housing benefits, disability benefits, and all the rest. It's one thing for a Green Party to propose a hippy something-for-nothing policy; perhaps the stern right-wing efficiency case ought to be made more often.

As it stands, the current benefits system, Jacobson points out, is a classic trap, in that it doesn't pay to start working if your benefits are taken away nearly as quickly as you earn extra money. 'The marginal tax rate on most means-tested benefits,' she explains, 'is anywhere between seventy-five and ninety-eight per cent, in terms of loss of benefit plus tax on your income. Governments don't seem to recognise what a huge disincentive to work that is.' She recalls talking to a tax accountant who was explaining this problem to businesses: '"The reason you can't get a receptionist for four hours a day for two days a week," he said, "is that it's not worth their while, it's over the amount of time that they can work according to their benefits, so it could actually lose them more money than it gains them in the end." Whereas if you had a Basic Income they could work quite happily, you just work on top of that.'

Not only is the current system a trap, it can actually be bad for claimants' health. Jacobson herself spends her days helping people through the bureaucratic maze of the Employment and Support Allowance. For those claimaints 'who have mental health problems', she says, a UBI would 'just be astonishing'. Furthermore, 'if it was high enough, then people with disabilities wouldn't have to be emphasising how sick they were all the time; they could actually develop and do things and start emphasising what they can do. Whereas again, at the moment, it's "I've got to prove how sick I am." People's health, you can watch it deteriorate as they go through all these processes to get their benefits. It's absolutely outrageous.'

In sum, Basic Income should not be ghettoised as a utopian leftist idea. 'Any thinking conservative can see that poverty costs the country a lot of money,' Jacobson says. 'Whether that's in terms of crime, or health problems, or lost days at work. So there is not just the anti-bureaucratic case for Basic Income, but also the benefits-to-the-public-purse case, that conservatives – in the past, anyway – have signed up to.'

Your kicks for free

I wonder whether the resistance the idea still meets in some quarters is because it somehow offends a fundamental sense of fairness. Do we have a reflex moral reluctance to give everybody something for nothing? One answer, of course, is that the rich already get a lot of money for nothing (interest, rents, returns on investments, and so on), so why shouldn't the rest of us?

One might also respond that it's not 'for nothing' in any case; rather, as Paine and Fourier argued, it's an acknowledgement of our shared common inheritance and the restrictions on our rights that civilisation insists on. Another retort is offered by Cunnen: 'Corporations are already getting something for nothing, but nobody talks about corporate welfare.' (The system of working tax credit in the UK, for example, uses public funds to top up workers' insufficient salaries.) Jacobson agrees: 'Exactly. Some people, particularly on the left, argue that Basic Income itself would then be a subsidy to employers, but if you set it high enough then it actually gives people the power to say no.' And so employers would be forced to pay more. 'What we call a job market is not actually a market at all,' Jacobson says, 'because the people who are offering their labour are being forced to work or they'll starve. That's not really a free market.'

Another answer to qualms about fairness might be to alter the source of funding. Jacobson says: 'There's a real problem in the fact that benefits come from income tax or National Insurance, and so it's basically a tax on slightly-less-poor people and then it goes to poor people.' This tends to breed social resentment, of the kind that is easily fanned by tabloids screaming about benefits fraudsters and so on. Preferable, for Jacobson, would be a situation in which a UBI or other benefits would be coming 'from profits, or from unearned income, whether that's rents or higher inheritance taxes'.

Another idea, proposed by the economist and professor of development studies Guy Standing, is that money from copyright and patents that exceeds a certain level should go into a sovereign wealth fund. 'Because all invention and all creativity,' Jacobson points out, 'is actually a product of our ancestral heritage as well as the people around us.

'It's not just one person sitting in a room,' she adds, 'that invents anything or thinks of anything, really, or creates anything.' As Paul Feyerabend said: no invention is ever made in isolation. Yes, the kind of businessman who boasts of being 'self-made' would object, but Steve Jobs did not build the roads on which Apple computers could be transported to shops. A hedge-fund manager did not pay for the mathematical education of his formula-wielding money wizards. Thinking and rethinking are irreducibly collaborative. Even if your collaborators are just long-dead writers.

Too simple to be true?

Another source of resistance to Basic Income might be the idea that it is just too simple. Problems in the real world are complicated, and you can't solve complicated problems with such naïve solutions. In such debates we often use the phrase 'just throwing money at the problem', which is understood to be the kind of unsophisticated idea that would have all kinds of counterproductive consequences.

In fact, as it turns out, just throwing money at the problem is often the best way to solve it. Especially if the problem in question is that people don't have enough money.

In the past few years, observers have been mysteriously amazed that projects in Brazil, Uganda, Kenya and elsewhere which just *give money to poor people* seem to be the most effective way yet of combating poverty. It's another radical idea. And it works. Even the bastion of free-market economics, *The Economist* magazine, admitted: 'Giving money directly to poor people works surprisingly well.'[19] The only surprise is that anyone would be surprised.

In Kenya the Justgiving organisation is experimenting with providing Basic Income to whole communities. 'That seems to really be far more effective in alleviating poverty than anything else,' Jacobson says. Just give them money, huh? 'It's very simple, yeah!' Cunnen says. 'I go to conferences on benefits, with benefits advisors, and talk about Basic Income,' Jacobson says, and everybody goes: "Oh no! That's too simple! You can't just solve these things!"' But maybe you can. It just requires a bit of psychological adjustment.

Naturally, we tend to value complex and sophisticated ideas. Ideas that seem too simple we call naïve, because they don't take account of all the nuances and shades of grey in the world. But sometimes, a really simple idea – like *just giving money to people* – is a really good one.

The only fly in the ointment, it seems to me, is that the phrase 'Basic Income' itself is rather uninspiring. 'Basic' is not the most thrilling word, is it? Jacobson and Cunnen laugh. Other terms for the same idea currently include 'Citizen's Dividend', 'Social Dividend', 'Social Wage', or 'Guaranteed Livable Income'. In the nineteenth century, Joseph Charlier named his revolutionary proposal the 'Territorial Dividend'. Which to choose? 'I really like the term Social Dividend,' Jacobson says, 'because Basic Income has been used so much in the last thirty years, and people get it mixed up with minimum wage.' 'And living wage,' Cunnen interjects. What's more, with the word 'dividend', Jacobson says, 'People don't automatically assume you should work for it. So we may start using that a bit more.'

There we go. Social Dividend. This is evidently an example of Unspeak: a political label that smuggles in an argument in the idea's favour. After all, the term 'social' is a fashionable marker for warm and fuzzy feelings (compare 'social media'), and 'dividend' cleverly takes a concept from speculative finance (a shareholders' dividend) and democratises it.[20] At least it's all in a good cause.

But who is going to take the plunge first? Which country will actually institute a Social Dividend, and how soon? We know it is high on the agenda in Switzerland. And in Finland's general election of April 2015, all the major political parties came out in support of the idea, with the newly formed government committing to run a pilot project scheduled to begin in 2017. For such an apparently idyllic idea, it seems as though reality is finally getting close. 'I mean, Virginia Woolf was calling for it in *A Room of One's Own*,' Jacobson points out. 'What we all need to discover "Shakespeare's sister", she said, is £500 a year – which translates into £20,000 now – and a room of one's own, you know?' That was in 1929. Maybe, by the centenary of Woolf's essay, Jacobson suggests, 'We'll find Shakespeare's sister again.'

But to certain minds the only way to accelerate the adoption of such reasonable and egalitarian proposals would be to reform the political class altogether. We live in an age of the professionalised politician who seeks re-election before all else, and the range of policies offered by the main parties is dramatically limited. When the comedian and actor Russell Brand told the youth of Britain not to bother voting in 2013, he was playing on widespread recognition – in the political-philosophy literature, as well among ordinary people – that modern representative political democracy is broken. Governments are captured by wealthy lobby interests at home, and in thrall to speculators in the bond markets abroad. The professionalised political class has little to no experience of normal working life and cares only for the perpetuation of its own power. In order to secure re-election it must assuage rich benefactors, and the four- or five-year cycle of elections renders it incapable of taking the longer view required to act in society's best interests given long-term challenges such as global warming. Mitt Romney's secretly recorded boast, during the US presidential campaign of 2012, that he could afford to ignore the wishes of 43 per cent of Americans, was just an unusually honest admission of the facts. That is how our 'democracies' work.

Perhaps politicians themselves are the problem. It already seemed that way to many people more than a century ago. One of them was the English novelist Thomas Hardy. In 1885, after Gladstone stepped down as prime minister, Hardy wrote in his diary that, while doing the rounds of London dinner parties in the preceding weeks, he had experienced no better than 'intensely average' political conversation – and that from the lamentably 'average men' he encountered who were soon afterwards installed in the Cabinet. 'A row of shopkeepers in Oxford Street taken just as they came would conduct the affairs of the nation as ably as these,' Hardy moaned.[21]

Thomas Hardy's complaint was meant to be an insult to politicians rather than a declaration of faith in shopkeepers. But what if we took it seriously? What if we were governed, not by career politicians, but by a randomly chosen assortment of our peers?

This, too, is a very old idea.

The national lottery

In many modern Western societies people can be called up for jury service. They are chosen at random, so that the average jury will represent a cross-section of society. Perhaps a couple of them will be terrible idiots, but it's very unlikely that all of them will be. Personal foibles will be balanced out in the discussion among twelve individuals. Overall the jury has been trusted for centuries to make reasonable decisions based on the evidence when deciding the fate of their fellow citizens. Of course there are miscarriages of justice and decisions that others consider perverse, but these may well happen less often than under rival systems such as those in which guilt is determined by inquisitorial judges. Indeed, trial by jury is arguably one of the best ideas in history.

Why not, then, apply the same principle to society at large? Instead of a professionalised class of governors, simply choose leaders at random from all ordinary walks of life. They will be well paid and serve a fixed single term. Perhaps many of them will be terrible idiots, but so are many people who manage to get themselves elected under the current system. And it's highly unlikely that several hundred people chosen at random would *all* be terrible idiots. Such a 'jury government' could very well be more in touch with normal people, and it would certainly be free from the electoral and lobbying pressures that create short-termism, superficiality, and corruption.

Sure, it sounds fanciful. But that is how the ancient Athenians did it. The inventors of democracy themselves chose their government by lottery. It's called sortition, and maybe it's time to try it again.

In classical Athens, selection by lottery was used for three out of four major political institutions. And the system persisted as recently as the early Renaissance, in Italy. Indeed, 'For most of European history,' the anthropologist David Graeber explains, 'elections were assumed to be not a democratic, but an aristocratic mode of selecting public officials. "Aristocracy" after all literally means "rule by the best", and elections were seen as meaning that the only role of ordinary citizens was to decide which, among the "best" citizens, was to be considered best of all [. . .] The democratic way of selecting officials, at least from Greek times onwards, was in contrast assumed to be sortition.'[22]

How exactly would sortition work in the modern age? The question is exercising some serious political thinkers. The philosopher Alexander Guerrero calls his own model 'lottocracy'. Its main feature is that, rather than having one big assembly, it has a couple of dozen 'single-issue legislatures' – one for agriculture, one for health care, one for education, one for transport, and so on. The advantage of this, Guerrero points out, is that citizens serving on such an expert-advised legislature for several years would have much more time to learn about the topic than do current politicians, who can be reshuffled at a moment's notice and suddenly have to pretend to be expert in education when the week before they were expert in health.

'People would not be required to serve if selected,' Guerrero writes, 'but the financial incentive would be significant, efforts would be made to accommodate family and work schedules, and the civic culture might need to be developed so that serving is seen as a significant civic duty and honour.'[23] It might on the whole be better, however, if serving were mandatory with exemptions made only for very significant counter-commitments (say, caring for a sick family member). After all, people currently say that a problem with jury service is that middle-class professionals can exempt themselves too easily.

Sortition also enables governors to think about everyone, rather than feeling hidebound by a duty to represent the special interests of the particular local area that elected them in the first place. 'Instead,' Guerrero says, 'they will be like better-informed versions of ourselves, coming from backgrounds like ours, but with the opportunity to learn and deliberate about the specific topic at hand.'

Sortition, too, sounds very simple. It is a fundamental, radical rethinking of how government should work. The problem is the entrenched interests it faces. Just as turkeys don't vote for Christmas, it is difficult to see the entire political class voting itself out of existence. But there are more of us than there are of them.

The return of Utopia

Sortition and Basic Income (or Social Dividend) are both often described as 'utopian'. Utopianism in our age is widely reckoned to be exclusively a leftist, progressive disease. The horrors of Stalinism, it is often said, show all too clearly where utopianism can lead: to totalitarianism.

But there is another kind of utopianism at large that is not often labelled as such. It is the opposite of totalitarianism because it is suspicious of any kind of human planning. It is the faith that free markets, with minimal or no regulation, can best provide all social goods. (Sometimes this is called neoliberalism.) Its adherents sincerely believe that any kind of top-down thinking to improve society ought to be discouraged, because that way lies the Gulag. Instead, the operation of the free market will, without any planning or direction, produce the best possible outcomes for all. Well, will it? The last decade of global economic history doesn't seem to support that belief.

I once heard a conference talk by the philosopher Jamie Whyte. Central government planning of school curricula, he said, was utopian and doomed to failure. Instead, he argued, there should be an entirely free market in primary and secondary education, with schools able to teach anything they like. After a while (though he didn't say how long), all the schools teaching rubbish would have failed, and only the good schools providing a fine education would be left. Voilà, unplanned perfection.

At the end of his talk, I suggested to Whyte that his own idea was itself utopian. He seemed quite sanguine, after all, about the prospect of ruining the education of many children now (those who would be sent to the schools which turned out to be teaching garbage), in the service of an unproven, untimetabled future when the garbage-teaching schools would all have failed. Whyte's response was merely to say that children's education was, in his view, already bad under the national curriculum. But does that really mean it's all right to make it even worse? Those who think this is worth doing – another is the zoologist Matt Ridley, in his book *The Evolution of Everything* – are possibly quite well described with a phrase that the conservative philosopher Roger Scruton has applied to liberals: 'unscrupulous optimists in their war against reality'.[24]

The idea that competition will eventually result in the best possible schooling is utopianism – but utopianism based on sheer faith rather than forethought and planning. We don't have a choice, here, between utopianism and modest reasonableness. We must choose between two utopias. And it's not at all clear that the totally unplanned one is necessarily better than the one where people think carefully about things. In any case, perhaps the critique of progressive proposals as utopian has lost its sting. Perhaps the rethinking of political economy since the global financial crisis has ushered us into a neo-utopian age.

Nothing wrong with that, if so. Writing of sortition, Alexander Guerrero ends on an inspiring note: 'This task of radically redesigning government is usually dismissed as utopianism, but there is no reason to think that electoral representative democracy can't be improved upon, just like every other kind of technology.'[25] Well, sure. If we can make better electric cars and better anti-malarial drugs, surely it is not beyond our ability to make better political systems. All it takes is a little rethink.

Why not?

Why don't we just give everyone enough money to live on? Well, that would never work: they'd just laze around doing nothing. Why don't we just give money to poor people? Well, that would never work: you can't just *throw money at the problem*. Why don't we choose our politicians by lottery? Well, that would never work: ordinary people can't possibly be trusted to run the country . . . But actually these things do work – in trials conducted in towns and villages, and in ancient societies. The experiments have been done. The results are in. To the extent that both basic income and sortition are also examples of power-up ideas – they will, like Grace Hopper's vision of programming, empower more people beyond the current elite – it is of course understandable that they meet resistance.

In politics as in other fields, to rethink is to keep asking 'Why not?', like an unnervingly persistent child unwilling to be fobbed off with hasty and bad explanations. Maybe it helps to rework the framing of an idea – using the phrase 'Social Dividend', for example, instead of

'Basic Income'. If we are accused of utopianism we'll say: Sure, but aren't you too? Isn't the prevailing view already utopian? We always have a choice between utopias.

The fear of what might go wrong if we carefully plan our utopias, rather than leaving them to the unpredictable operations of a market, is really pessimism about the human power of collective reasoning. And so it is a counsel of despair. If people working together can design a space programme that sends humans to the Moon and robots to Mars, it doesn't seem that difficult for people working together to design a school curriculum that will give the best chance of a decent education to everyone. Or, for that matter, to build a better system of government. If you are indoors or in an urban environment right now, take a moment to look around you. Every non-living object you see was thought up, designed, and made by human beings. That is *amazing*. Of course, limitless faith in the rational capacity of co-operating humans might be misplaced. But if we didn't have any such faith, we'd never get anything done. It's an idea that is important and useful, even if it might be inaccurate. It's another placebo idea to live by.

In the mean time, the fact that many people are arguing for the modern viability of old ideas like Basic Income and sortition shows yet another way in which our age is not so novel and unprecedented that past thinking can't be productively applied to it. As the economist Joseph Stiglitz remarked when recommending a renewal of New Deal policies for present-day America: 'Just because you've heard it before doesn't mean we shouldn't try it again.'[26]

12

Beyond Good and Evil

What evil ideas from the past might be worth another look? And which ideas of ours will look horrific to our descendants?

'Ages are no more infallible than individuals; every age having held many opinions which subsequent ages have deemed not only false but absurd; and it is as certain that many opinions, now general, will be rejected by future ages, as it is that many, once general, are rejected by the present.' – J. S. Mill

The huge new Francis Crick Institute for molecular biology has an undulating, convex metal roof that makes it resemble a giant steel-backed armadillo, crouched behind St Pancras in London. The scientists have not yet moved in to the building when, one autumn afternoon, I wander into the Visitor Centre. It has a small exhibition featuring reassuring photographs of smiling scientists helping to cure diseases. When the project was first mooted years ago, one of the institute's friendly publicity officials tells me, local residents were worried. 'There were rumours and a bit of scaremongering going on, like "What *are* they going to study in there?"' Later, neighbours were nervous about potential Ebola outbreaks, and thought that there might be a related terrorist threat. Institute officials calmed their fears at community meetings.

And then, in September 2015, the news came out that scientists from the Crick Institute had applied for permission to conduct research involving the genetic engineering of human embryos. The media coverage got people talking again about what dastardly science might eventually be conducted within the armadillo.

'Oh yes,' the other publicist says with a tolerant smile. '"Designer babies" came up, as usual. And we said, "Well, this project's actually currently approved for just a two-week period, and it just involves a slight modification", but never mind that . . .' She laughs, like someone used to calming irrational panics.

But what's so wrong with designer babies, anyway?

Pariah ideas

Some ideas seem to be right or wrong in a moral sense, and in a timeless, objective way. But we also know that ideas have been accorded varying moral evaluations in different contexts through history. Moral attitudes can shift quite rapidly, indeed, during a single human lifetime: an obvious example in recent years has been the shift in views about the idea that gay people could get married. So rethinking can involve taking an evil idea that lurks in the past and giving it a dispassionate re-evaluation so that it comes to seem neutral, or even good.

To observe that moral attitudes towards ideas and social practices change over time and across different societies is often to invite the indignant accusation of *moral relativism*. But it's an incontrovertible truth. The age of sexual consent, for example, has been widely different at different times and places throughout human history. It has often been common for adult men to marry girls of twelve or thirteen. To insulate oneself entirely from the terrible charge of moral relativism, it would be necessary to arrive at a universal judgement about what is right and wrong on this subject for all time. Then, by implication, entire societies and cultures would have to be denounced as immoral. It might after all be more practical to limit ourselves to a judgement about what we, in our society, think is right for us right now. As for other arrangements – well, *autres temps, autres moeurs*. And the humane wisdom of that saying holds just as true if the *autres temps* are in the future. Our moral views will one day be rethought, just as those of past ages have been.

Certain pessimistic commentators on human history claim not to believe in progress. The Enlightenment, they say, just led to new rationales for mass killing. As it happens, people have been denying progress about as long as others have been celebrating it. Their modern successors do it with the benefit of electricity and good dentistry, but what they do is the same. There is no progress to be expected, perhaps, in the field of denying progress.

Some will admit that technical progress has been made (electricity, dentistry), but they will reject any possibility of moral progress. Well, a century ago it was common and unremarkable for rich men to go around Africa shooting large animals and bringing home their heads to mount on their wood-panelled study walls. In 2015, when an American dentist shot a lion named Cecil that had been lured out of a nature reserve in Zimbabwe, there was widespread revulsion. That is surely progress for admirers of big cats. You might think, too, that trial by jury, the abolition of slavery, women's suffrage, universal education, and other things seem to constitute quite decent moral progress, at least in those countries lucky enough to have instituted them. Yes there are, as we have seen, moral zombies – the return of slavery, torture, and so forth in various places – but the existence of moral zombies does not negate the existence of moral progress generally. As the sci-fi writer William Gibson said: 'The future is already here; it's just not very evenly distributed.'[1]

Sometimes an idea is so strongly associated with past horrors that it becomes a kind of byword for evil. When it threatens to return, people invoke those past horrors again to try to prevent it even from being discussed. But this can be unjust. An idea might be unfairly tainted by a particular context in which it was used, and become a *pariah idea*. This may be the case, for example, with the idea of designer babies.

Fitter, happier

Let's say you have a child. Imagine that there is a freely available pill that is not very expensive and perfectly safe, which will enhance your child's intelligence by some significant amount. Do you give your child the pill? If we stipulate for the sake of argument that increased intelligence will, other things being equal, improve your child's life chances, why wouldn't you give her the pill? Giving the pill to your child seems comparable with lots of other normal parental activities such as paying for music lessons, or just helping her with her homework. You might lament the fact that there is now a kind of pharmaceutical arms race in which parents try to do better by their children than everyone else – but there has long been a comparable kind of cultural arms race. And this (imagine) is the world we live in now: where a cheap, safe pill can make your child's life better. In this context, it seems quite tempting to say that it would be wrong *not* to give your child the pill.

Now imagine something slightly different. A woman discovers she is newly pregnant. It so happens that there is a freely available pill that is not very expensive and perfectly safe. If the woman takes it, the pill will alter the genome of her foetus, so that later, when the child is born and develops, its intelligence will be enhanced by some significant amount. Should the woman take the pill, and so genetically engineer her baby to be cleverer? Indeed, by the same arguments as given in the first example, wouldn't it be wrong for her *not* to take the pill?

No doubt you've seen what I did here. And yes, it's true that answering positively to the second scenario would be supporting eugenics. But what is the morally relevant difference between pharmaceutical enhancement, in the first example, and genetic enhancement? What exactly is bad about eugenics?

It would seem as though the question hardly needs asking. The history of eugenics is littered with horrors – the forcible sterilisation of, the outlawing of marriage for, and often the outright murder (euphemised as 'euthanasia') of those deemed 'feeble-minded' or otherwise inferior in 'blood'. All this proceeded apace in the United States in the early decades of the twentieth century, and indeed it is thought that American eugenics influenced Nazi practice. In 1911 the Carnegie Institute had recommended 'euthanasia' as a solution to the supposed genetic deterioration of society, and the use of gas chambers was widely mooted, though it was decided the public would balk at such a programme and instead many doctors practised killing of the 'defective' by 'lethal neglect'.[2] In 1936, an American eugenicist named Harry H. Laughlin was awarded an honorary doctorate from Heidelberg University in Germany for his contribution to the 'science of racial cleansing'. Eugenic ideas had been quite common, too, among British and European socialists and progressives. And then, after the giant Nazi programme of murder during the Second World War, eugenics could hardly be spoken of again unless it was to hold it up as a terrible example of the hell to which pseudoscience can lead.

So it went. But the fact that an idea *has* been used for evil purposes does not entail that it will *always* lead to evil. Even so, that is an all-too-common slippage. Most people recoiled from the idea of eugenics, one modern journalist writes, 'once they saw where it led – to the gates of Auschwitz'.[3] It is perhaps fair to ask: did it lead there, or was it pushed? Are some ideas evil for all time? Or can an idea be evil one day and good the next?

Let's divide the issue into two. Call 'negative eugenics' the practice of not allowing to develop to maturity (or, historically, killing or badly disabling) potential human beings who would have qualities we judge to be detrimental. On the other hand, let's say 'positive eugenics' is the engineering into potential human beings of traits we judge to be desirable. (In the heyday of eugenic thinking, the latter term was used slightly differently: 'positive eugenics' meant encouraging 'higher-quality' people to have more children. Of course, back then, engineering specific traits into children was not possible. Today, calls for one or another group to breed more are limited to the fringe of racist culture-war.)

Negative eugenics is the source of most of the bad rap that eugenics has had. Note that it is still in a way practised in our time, through the termination of foetuses that it is judged would develop into children whose lives would consist of unremitting suffering. Still, most people would agree that it would be wrong to extend the scope of negative eugenics, say by terminating foetuses because the resulting children would not be very good at sport.

It is more difficult, on the other hand, to make a case against modern positive eugenics. In any case, people already practise it in an amateurish and haphazard fashion, simply through their choice of breeding partners. (This man is tall and funny, therefore our children might have a better chance of being amusing and statuesque as well.) As techniques of genome manipulation come to maturity in the future, why not systematise things so that prospective parents have a better idea of what they're getting? This is a debate that needs to happen soon since the technological possibilities are rapidly catching up with science fiction. The new gene-editing technique called CRISPR ('clustered regularly-interspaced short palindromic repeats') was hailed by the journal *Science* as the most important breakthrough of 2015. It's a bit like a word processor for DNA. Researchers using CRISPR have already made the first successful steps towards reversing blindness caused by genetic disease in adult patients.[4] And the CRISPR technique can also modify the DNA in human embryos. It may therefore represent the missing piece that renders eugenics not only practicable but – in some cases – morally obligatory.

One argument that might tell against this idea in practice is that eugenic interventions would presumably be costly and not equally available to all – so that the descendants of the wealthy would become ever more beautiful and clever, while the descendants of the poor would be condemned to the genetic lottery and fall ever further behind. Likely as this may be in the context of modern capitalist elective medicine, it does not in itself count as an argument against positive eugenics in principle. It is simply a warning that the political mechanisms of its availability need to be carefully administered in case we don't want to end up in a two-tier society. (In many other ways, we do in any case.)

More to the point, perhaps, is the question: well, who's to say what counts as 'improvement'? It might be that, thanks to the widespread availability of aesthetic eugenics, a greater proportion of the population comes to be blond-haired and blue-eyed. This would be good news for admirers of the Aryan ideal, but those who appreciate brunettes would not necessarily count it a global upgrade.

Still, perhaps it is time to think seriously again about an enlightened eugenics. Not only because biotechnological modifications are on the way, and we can't stop them. But perhaps because wholesale improvement of the human race really would be – well, an improvement. We might soon have the technology to rethink humanity itself. Should we use it?

The father of eugenics

Francis Galton, half-cousin to Charles Darwin, was another great Victorian polymath. He invented important concepts in statistics; he classified the patterns in fingerprints, leading to their acceptance as evidence in the courts; he prepared the first-ever newspaper weather map; he founded the discipline of psychometrics (using questionnaires to glean information on people's mental faculties) – and he was a eugenicist. Indeed, he coined the word 'eugenics', a Greek construction literally meaning 'good inheritance'. (His first attempt at naming what he called 'the science of improving stock' was the less catchy 'viriculture'.)[5] And for that his name, certainly in the latter half of the twentieth century, came to be widely despised.

Yet Galton's project was in some ways a humane one. The idea was that rational humans could now take charge of their own development, with less horrific cruelty than nature employs. 'The process of evolution,' he pointed out in 1883, 'has been hitherto apparently carried out with, what we should reckon in our ways of carrying out projects, great waste of opportunity and of life, and with little if any consideration for individual mischance. Measured by our criterion of intelligence and mercy, which consists in the achievement of result without waste of time or opportunity, without unnecessary pain, and with equitable allowance for pure mistake, the process of evolution on this earth, so far as we can judge, has been carried out neither with intelligence nor truth.'[6]

In other words, evolution on Earth had hitherto proceeded through the blind operations of natural selection, which can operate only by means of suffering and death. (The slogan 'survival of the fittest', coined by Herbert Spencer in 1864, tends to obscure the equally valid perspective that it is the non-survival of the non-fittest which is the true, bloody engine of evolution.) But then along came conscious and reasoning minds in the form of human beings, the appearance of which seemed as astonishing to Galton as it did to Thomas Nagel more than a century later. ('Our personalities may be the transient but essential elements of an immortal and cosmic mind,' Galton speculated prettily.)[7]

Now that human beings exist, Galton thought, it is within our power to be less monstrously cruel than nature itself – and since it is within our power, it is actually our obligation: 'Man has already furthered evolution very considerably, half unconsciously, and for his own personal advantages, but he has not yet risen to the conviction that it is his religious duty to do so deliberately and systematically.'[8]

So far, so reasonable. The trouble to modern ears comes, however, when Galton begins to speak confidently of the necessity for the 'low' or 'inferior' races of humankind to vanish. Not by force, he argues, but simply by encouraging the 'higher' races to breed more. 'The most merciful form of what I ventured to call "eugenics" would consist in watching for the indications of superior strains or races, and in so favouring them that their progeny shall outnumber and gradually replace that of the old one.'[9] He knows that this might sound bad: 'There exists a sentiment, for the most part quite unreasonable, against the gradual extinction of an inferior race.' But, he says, this sentiment 'rests on some confusion between the race and the individual, as if the destruction of a race was equivalent to the destruction of a large number of men. It is nothing of the kind when the process of extinction works silently and slowly through the earlier marriage of members of the superior race, through their greater vitality under equal stress, through their better chances of getting a livelihood, or through their prepotency in mixed marriages.'[10]

What we can say of Galton, then, is that he accepted the pseudoscientific racism of his time – as did Darwin and almost everyone else –

but not that he was an exterminationist. Indeed, his particular historical misfortune is arguably that he did not specifically rule out any methods to achieve eugenic goals. Considering the objection that, 'until some working plan is suggested, the consideration of improving the human race is Utopian', he decided that the requirement for him to formulate a working plan was not fair: 'because if a persuasion of the importance of any end takes possession of men's minds, sooner or later means are found by which that end is carried into effect'.[11] Well, sooner or later they were. And they involved forced sterilisation and mass murder.

Bioethics

So Galton in the later twentieth century became an intellectual pariah, and eugenics became a pariah idea – tainted with guilt by association. For this reason many people who champion the modern possibilities of bioengineering avoid using the term 'eugenics', preferring to speak of 'human enhancement'. (Like 'Social Dividend', this is arguably an example of argumentative Unspeak. Who, after all, could possibly be against 'enhancement', especially enhancement of all humans?) Others think that we should face squarely up to the fact that such developments would be eugenics, and boldly recuperate the term itself. This school of thought often goes under the banner 'liberal eugenics', which sounds pleasingly paradoxical and so is a decent attempt to spark curiosity.

Modern arguments against enhancement, or liberal eugenics, do not fare well when they simply invoke Galton and the Nazis, because modern positive eugenics nowhere suggests that some people should be prevented from having children, still less that they should be killed. Nor need we give much consideration here to fears about 'playing God', since those who play the playing-God card rarely agree that we should also stop trying to play God by breeding edible crops or developing antibiotic medicines.

Are there, then, any good arguments against liberal eugenics? A 2014 book recording the results of discussions at the Scottish Council for Human Bioethics is a useful overview of the main arguments pro and con. The contributors finally argue against it, on the grounds that to accept eugenic selection for its possible improvements to an individual's quality of life is to deny the 'inherent dignity and worth' of all people, as set out in the UN's Universal Declaration of Human Rights.[12] It is not clear, however, that 'inherent worth' and selection of attributes for a certain conception of 'quality of life' are really opposed in this way. If we value a person in one way, are we refusing to value them in the other? Consider an adult's selection of a sexual partner. If a woman chooses to sleep with this guy rather than that guy, she is not thereby necessarily making any negative objective judgement about the 'inherent worth' of that guy. She just prefers this guy. Similarly, if a parent chooses to engineer her child to have athletic skill, she is not thereby announcing that unathletic maths geniuses are of lesser inherent worth.

If the idea of 'quality of life' is used in such a way as to imply that some life has *zero* quality and so, in the Nazi slogan, becomes 'life unworthy of life', then that is indeed troubling and ought to be resisted. But otherwise, the act of selection (of certain attributes in a child) does not seem fundamentally opposed to the idea of the inherent worth of all children. Declarations of the inherent dignity and equality of all human beings, after all, do not oblige us to refuse to distinguish between human beings on any grounds at all – reproductively, sexually, or just because we like some more than others. The principles of inherent dignity and equality protect people from unjust discrimination by states on grounds of ethnicity, wealth, and so forth – but they do not protect people from any kind of discrimination at all, such as the discrimination involved in being rejected for a job because one doesn't have any of the appropriate skills. We may, further, choose to ask: what about the inherent dignity of parents' right to choose what they think will be best for their families?

The most thoughtful challenge to eugenic freedoms, perhaps, is that of the German philosopher Jürgen Habermas. 'Exercising the power to dispose over the genetic predispositions of a future person,' Habermas writes, 'means that from that point on, each person, whether she has been genetically programmed or not, can regard her own genome as the consequence of a criticizable action or omission. The young person can call his designer to account, and demand a justification for why, in deciding on this or that genetic inheritance, the designer failed to choose athletic ability or musical talent,' and so forth.[13] Will children always be happy to have received the gifts their parents chose for them? Or will those abilities come to seem like burdens? That is a foreseeable interpersonal and social problem for positive eugenics. Whether it outweighs the other benefits is a matter for reasonable argument.

At least one benefit seems indisputable. As the co-discoverer of the structure of DNA, James Watson, put it: 'The genetic dice will continue to inflict cruel fates on all too many individuals and their families who do not deserve this damnation. Decency demands that someone must rescue them from genetic hells. If we don't play God, who will?'[14]

We would need an extremely strong justification to refuse to alleviate human suffering in this way. We would need a strong justification, too, to refuse to enhance human flourishing if it were possible. Remember, we already practise eugenics – just not very systematically. A further thought is prompted by the suggestion from neuroepigenetics that stress and depression arising from trauma can be passed down to the next generation. This implies that parents who avoid traumatic situations are thereby acting eugenically, protecting their children from an undesirable genetic inheritance. What if that could be done more precisely? If CRISPR-style genome editing becomes reliable and widely available, so that eugenics *can* be practised systematically, it will plainly alleviate a lot of suffering that would otherwise be caused by hereditary disease, and it may even lead to a population of more talented and attractive people. These, if attained, would be concrete benefits to be weighed against any harms, in debates about regulation. There are already laws in place that forbid certain procedures of genetic selection: in the UK, for example, it has been illegal since 2008 to positively select, during fertility treatment, an embryo that will have a disability. In such ways, certain *dysgenic* (the opposite of 'eugenic', meaning 'bad inheritance') interventions are outlawed. So we ban the choice of harm. Why not also allow the choice of help?

Reasonable people may disagree on the details and extent of modern eugenic procedures, and some may argue against them for sophisticated moral reasons such as those offered by Habermas. But the very fact that such arguments are being conducted today shows that eugenics is no longer largely unthinkable, as it was for most of the second half of the twentieth century. It is an idea that is slowly losing its pariah status.

Indeed, eugenics may be going through a historical transition period on the way to becoming positively desirable. In the future, that is, it may come to be considered morally wrong *not* to avail oneself of at least some eugenic technologies. Decency demands it, says James Watson. That, too, was the robust view of Robert G. Edwards, who won the 2010 Nobel Prize in Physiology or Medicine for his role in the creation of IVF treatment. In 1999, he said: 'Soon it will be a sin for parents to have a child that carries the heavy burden of genetic disease. We are entering a world where we have to consider the quality of our children.'[15]

Edwards makes a very strong point here. The modern ability in IVF treatment not only to identify known genetic defects in an embyro but then to *correct* them – so that, for example, the child will be born without sickle-cell disease – is a eugenic intervention, and one of the most majestic triumphs of modern medical science. It is hard to dispel the notion that the 'quality of life' of someone born without sickle-cell disease because of such treatment is higher than that of a less fortunate person who lives with the disease, and that anyone who had the chance would choose the former for their child – without thereby denying the inherent worth and dignity of all.

So if eugenics can begin as misplaced optimism, then motivate mass murder and become unthinkable, and then become so necessary that it is a 'sin' or a dereliction of moral duty to eschew it, then it will become obvious to all that its former pariah status was linked to a particular sociopolitical context. Even Nazi science – as is widely acknowledged today, though still sotto voce – contained some good ideas, despite its atrocities. There is the obvious example of the Nazis' pioneering work in rocketry (which prompted the US to bring over Wernher von Braun, creator of the V-2, after the war to work first on ballistic missiles and then at NASA). According to scholarly works such as Robert N. Proctor's *The Nazi War on Cancer*, the Nazis were also in some ways ahead of their time in cancer research and other aspects of public-health policy such as nutrition.[16] They are the most extreme possible case of the principle that you shouldn't necessarily reject the message just because you don't like the messenger.

Future shock

If bad ideas can become good, then good ideas can become bad. So which of the ideas we live by right now will come to seem evil in the future? Unless we think moral progress has now completed its job and we live in a perfectly just world, some of our practices and assumptions will surely, to our descendants in centuries to come, seem thoroughly depraved or idiotic. The question is, which ones? Which of your ideas will make you a pariah one day?

Some ruling ideas seem to be on the way to becoming pariahs already: they are the current status quo but they are subject to increasingly loud and vociferous criticism. One, for example, might be eating the flesh of animals. The more we learn about animals, the more sophisticated they become. Pigs are really quite intelligent; even fish feel pain. Strict vegetarians would probably balk at eating sociable insects, too. In the future, the time when human beings routinely ate animals will probably come to seem an era of shocking barbarism. And I write this as someone who really does not want to give up his steak, chicken, and pulled pork. (Yes, that is one plateful.) Luckily, I might not have to. If current developments in the field of lab-grown animal protein – or, as the slogan puts it, 'meat without feet' – result in our being able to produce cheap and tasty chicken breasts without ever having to grow the rest of the chicken and kill it, then all will be well. Of course some people have been urging vegetarianism on us for centuries, so this is not a new idea, but it seems likely that in the future the idea of eating animals slaughtered for that purpose will acquire a taint of evil that for most people it doesn't today.

Another idea that is becoming a pariah before our very eyes is our current attitude to drugs. Remember the excitement around research into LSD for treating depression? Its advocates point out that LSD is not actually that dangerous. Yet it is still a Class A restricted substance. As the psychiatrist James Rucker pointed out in the *British Medical Journal* in 2015: 'No evidence shows that psychedelic drugs are habit forming; little evidence shows that they are harmful in controlled settings; and much historical evidence has shown that they could have use in common psychiatric disorders.' Yet it remains a condition of membership of the United Nations that countries prohibit many psychotropic drugs including LSD, 'stigmatising a facet of behaviour and arguably causing more harm than it prevents'.[17] The same goes, in many people's view, for the prohibition of *all* currently controlled substances. Pointing to the therapeutic benefits of LSD or psilocybin is a compelling argument for changing the legal status of *some* currently prohibited drugs, but not for legalising all of them. The reason why you'd cease to make it illegal to take any drugs at all is at bottom very simple: far less harm to far fewer people would result, compared to the status quo ante.

One day, indeed, the War on Drugs will probably be seen to have been even more counter-productive and perversely misery-inducing than America's grand experiment with the prohibition of alcohol. Drug enforcement in the US alone has cost more than a trillion dollars since Nixon first announced the War on Drugs in 1971. The figures collated by Law Enforcement Against Prohibition, a group comprising police officers and federal agents, are startling.[18] Nearly half of America's prison population of more than two million is incarcerated for drug offences – in other words, an eighth of the entire prison population of the world is locked up in the United States for possessing or selling banned substances. (The way the drug war is currently prosecuted in the US is also racist. Nationwide, black Americans are much more likely to be arrested, and much more likely to be incarcerated, for drug offences than whites, despite similar levels of overall use.)[19] Yet none of this belligerent action on drugs has had the desired effect. Nearly half the American population admit to using illegal drugs, so the policy is not an effective deterrent. And drugs have become cheaper rather than more expensive.

There are several different kinds of argument for legalising the recreational use of drugs such as cocaine, heroin, LSD, and so forth, in the same way that cannabis is now legal in many states of the US. The grounds might be simply libertarian – what business is it of the government's what I put into my own body? Or one might compare the harm done by illegal drugs to the harm consequent on perfectly legal activities, like American football or horse-riding.[20] A third kind of argument for drug legalisation would focus on the overall benefits to society. Consider that the vast majority of home burglaries and street muggings are committed by drug addicts seeking money for their next fix. If those addicts could instead get safe drugs cheaply (or even for free from government pharmacies), then a huge amount of street crime would be instantly wiped out. Naturally, this would be a positive development for many more people than drug users themselves.

Whatever the reasons, public opinion is gradually shifting. In a 1998 analysis of surveys conducted over the previous twenty years, 78 per cent of Americans believed that anti-drug efforts had so far failed, but two-thirds believed that more should be spent on the same strategies of enforcement.[21] By 2014, according to a Pew survey, two-thirds of Americans thought that drug policy should focus more on providing treatment to heroin and cocaine users rather than prosecuting them. A similar proportion of respondents thought it was a good thing that many states had abandoned mandatory prison terms for non-violent drug offences.[22] And more than half of Americans now support the legalisation of marijuana.[23] Maybe the War on Drugs is, at last, slowly beginning to wind down. Is it time to declare a war on something else instead? How about a War on Meat, or even a War on Driving?

Hands off the steering wheel

Allowing people in advanced countries such as the US and Norway to buy guns already seems like a very bad idea to many people, though its global outlawing is hardly imminent. But guns don't kill very many people at all compared with cars. More than a million people die every single year because of road traffic accidents, and tens of millions are injured. Traffic accidents, mainly due to driver error, are the single highest cause of death globally among people aged between fifteen and twenty-nine.[24] Hitherto, people worried about the dangers of drink-driving, or driving while texting, or driving while talking on the phone (even with a hands-free phone system). In the not too distant future, we may well consider these worries trivially peripheral to the really big problem: allowing people to pilot their own speeding hunks of metal filled with highly flammable liquid in the first place. People driving cars? Now that *is* a bad idea.

At least, it is likely to seem so once most or all cars are robotically piloted, like those currently being trialled by Google and other manufacturers. Perhaps driverless cars will never be completely accident-free, but the super-fast reflexes and unwavering concentration of a computer will surely do better on average than fallible humans. Computerised cars will be able to drive faster and closer together, reducing congestion while also being safer. They'll drop you at your office then go and park themselves. What's not to like?

Well, for a start, as the mordant critic of computer-aided 'solutionism' Evgeny Morozov points out, the consequences for urban planning might be undesirable to some. 'Would self-driving cars result in inferior public transportation as more people took up driving?' he wonders.[25] This is the kind of unintended consequence (an 'unknown') that an imaginative probe can uncover.

Another problem a world of self-driving cars would have to wrestle with was pointed out by Gary Marcus, professor of psychology at New York University. Suppose you are in a self-driving car going across a narrow bridge, and a school bus full of children hurtles out of control towards you. There is no room for the vehicles to pass each other. Should the self-driving car take the decision to drive off the bridge and kill you in order to save the children?[26]

What Marcus's example demonstrates is the fact that driving a car is not simply a technical operation, of the sort that machines can do more efficiently. It is also a moral operation. If we let cars do the driving, we are outsourcing not only our motor control but also our moral judgement. Very likely we will consider the price worth paying, to save those million deaths per year. But at the very least we'll have a serious stake in knowing exactly how those cars' algorithms are programmed, and which moral choices are buried within them.

Family planning

In Sweden and many other countries, you need a licence to own a dog; but in most of the world anyone can have children. Is that such a good idea? It hasn't always been thought so in the past. And it might not in the future either.

In the volume on eugenics by the Scottish Council on Human Bioethics, we find it claimed that in 1963 Francis Crick, co-discoverer of the structure of DNA, had written in a book entitled *Man and the Future* that some people were not fit to be parents, and that there should be a system of official licensing for procreation. That might seem disappointing to Crick's admirers. London's cutting-edge new research centre for biology is called the Francis Crick Institute. Should it really have been named after a man who thought some people didn't deserve to have children?

It's always worth checking this kind of précis of someone who can't answer back. In this case one finds that, in fact, Crick did not say exactly what is attributed to him, nor did he write any book called *Man and the Future*. He did, however, discuss the issue at a conference. And the truth of what he and others said is much more interesting. In 1963, the scientific non-profit Ciba Foundation held a symposium on

'Man and His Future', in which luminaries in biology and other fields discussed human evolution and the possibility of improving it. Yes: they were speaking about eugenics. The meeting, in Portland Place in London, took place not long after the Cuban Missile Crisis. This seemed to give special urgency to the topic at hand. As the geneticist Joshua Lederberg remarked to his fellow attendees: 'I think that most of us here believe that the present population of the world is not intelligent enough to keep itself from being blown up, and we would like to make some provision for the future so that it will have a slightly better chance of avoiding this particular contingency.'[27]

Francis Crick did not himself deliver a talk at the meeting, but he participated in general discussion after two papers on positive eugenics: how it might be possible to increase the general level of intelligence (say) of the population without coercive means. (One of the papers, by Herman J. Muller, suggested a system of sperm banks from which women could choose apparently superior donors – which has, of course, since come to pass.)

So what did Crick say? He certainly raised a hot-button question. 'I want to concentrate on one particular issue,' he said. 'Do people have the right to have children at all?' Then he began thinking science-fictionally. 'It would not be very difficult,' he pointed out, 'for a government to put something into our food so that nobody could have children. Then possibly – and this is hypothetical – they could provide another chemical that would reverse the effect of the first, and only people licensed to bear children would be given this second chemical. This isn't so wild that we need not discuss it. Is it the general feeling that people do have the right to have children? This is taken for granted because it is part of Christian ethics, but in terms of humanist ethics I do not see why people should have the right to have children.'[28]

Some of the others present agreed with Crick on the basis that 'society' ought to be able to decide what is best for society as a whole, if necessary overruling the preferences of individuals. Others disagreed strongly. Dr Alex Comfort, a zoologist who would later become famous as the author of *The Joy of Sex*, said: 'I would think that it is more true to say that whether people have the right to produce children depends on the circumstances. What I am sure of is that no other persons have the right to prevent them, which is rather a different matter.'[29] Some present astutely questioned how we could tell, or who would decide, which people were more 'socially desirable' than others.

In other words, this was an honest intellectual argument between scientific peers rather than an authoritarian statement of one man's ideal policy. Crick himself emphasised that he was conducting a deliberately provocative thought experiment in order to raise the issue. 'I think what I have described is a bit extreme,' he said, right after describing his possible licensing system, and that it would never be 'socially acceptable', so there was no point trying to institute it. A milder possibility, he suggested, might be simply to fiddle with tax incentives to encourage more 'socially desirable' people to have more children.

Whether one sides with Crick or with Comfort, the proceedings of the conference make fascinating reading on a topic that comes up again, as we have seen, in today's arguments over liberal eugenics. One thing Crick emphasised that might fruitfully be reconsidered in our age, where rights are imagined as accruing to the private, atomised individual, is the following: 'I think if we can get across to people the idea that their children are not entirely their own business and that it is not a private matter, it would be an enormous step forward.'[30]

In any case, Crick did not in fact say that some people were not fit to be parents. Yet eugenics is such a trigger for moral disgust even today that thinkers who can be shown to have flirted with such ideas are held up in a kind of triumph of disapproval, even when what they said was not exactly what is now reported. So it has been with Francis Crick, and Francis Galton. And so it has long been with another long-traduced thinker, Thomas Malthus.

It is always an illuminating exercise in rethinking to go back and read what a pariah thinker actually said. Very often we'll find that the sinister reputation is unjustified. So what did Malthus really think?

The population question

Negative eugenics – the attempt to prevent those considered genetically undesirable from having children – still languishes definitively in pariah status. Yet one can imagine it being revived in some dystopian future circumstances of terrible scarcity, when it might indeed be considered a moral wrong to procreate if one lacks the resources to support children. Thomas Malthus – he of 'Malthusianism' – considered this a moral wrong already in 1798, when he published *An Essay on the Principle of Population*. The book rapidly became notorious. In the following years, Malthus was attacked by Byron, Shelley, Coleridge ('monstrous practical sophism'), Hazlitt ('illogical, crude and contradictory reasoning'), and later Marx ('this libel on the human race').[31] So why is he still often thought of as such a monster?

In fact Malthus never argued that people should be forcibly prevented from breeding. But he thought (as Crick later did) that the social incentives were set up all wrong. He began by arguing that, when unchecked, human population tends to grow geometrically (by multiples), while the ability of farmland to support a population can grow only arithmetically (by additions). Because of this tension between human numbers and available food, an increase of population 'tends to subject the lower classes of the society to distress and to prevent any great permanent amelioration of their condition'.[32] If epidemic starvation and misery among the poor is to be avoided, then, 'checks' to population must be encouraged – primarily, Malthus argued, the postponement of marriage when resources are scarce.

Why is this a good idea? Malthus explains, quite sympathetically, the reasons that should motivate people in this way, giving examples from two strata of society. First he considers 'a man of liberal education, but with an income only just sufficient to enable him to associate in the rank of gentlemen'. Such a man, Malthus writes, 'must feel absolutely certain that if he marries and has a family he shall be obliged, if he mixes at all in society, to rank himself with moderate farmers and the lower class of tradesmen'. This is an age, of course, in which genteel women did not go out to work, so our liberal gentleman would have to divide his only-just-enough income to support a whole family, and it would no longer be sufficient to associate with the 'right' class. This, Malthus thinks, would be terribly unfair on his beloved as well. 'Can a man consent to place the object of his affection in a situation so discordant, probably, to her tastes and inclinations?'[33]

Next Malthus explains how similar reasoning should also discourage an honest working-class man from starting a family if he can't afford it:

> The labourer who earns eighteen pence a day and lives with some degree of comfort as a single man will hesitate a little before he divides that pittance among four or five, which seems to be but just sufficient for one. Harder fare and harder labour he would submit to for the sake of living with the woman that he loves, but he must feel conscious, if he thinks at all, that should he have a large family, and any ill luck whatever, no degree of frugality, no possible exertion of his manual strength could preserve him from the heart-rending sensation of seeing his children starve, or of forfeiting his independence, and being obliged to the parish for their support.[34]

All too many people, Malthus thought, were already 'obliged to the parish', receiving subsistence as mandated by the Poor Laws. Indeed, at the time more than 10 per cent of England's population of 10.6 million were on what we would now call 'welfare' or benefits.[35] This Malthus called 'dependent poverty', and he took a very dim view of it – for, as he argued, the recipients' own good. 'Hard as it may appear in individual instances, dependent poverty ought to be held disgraceful,' he wrote sadly. 'If men are induced to marry from a prospect of parish provision, with little or no chance of maintaining their families in independence, they are not only unjustly tempted to bring unhappiness and dependence upon themselves and children, but they are tempted, without knowing it, to injure all in the same class with themselves. A labourer who marries without being able to support a family may in some respects be considered as an enemy to all his fellow-labourers.'[36] To the objection that, in modern language, there is an inalienable human right to have children, Malthus would simply counterpose the right of children not to be born into poverty and misery.

From an unprejudiced modern point of view, then, Malthus appears no worse than many conservatives in the US or Britain, and indeed more genuinely concerned for the flourishing of his fellow human beings than many. (If you read the whole of his *Essay on Population*, it is difficult to doubt Malthus's sincerity on this point.) It is not even necessarily a right-wing position to critique the skewed incentives of a welfare system – recall Barbara Jacobson's view of the 'benefits trap'. To modern eyes, indeed, it may just seem a pity that for Malthus employing artificial methods of contraception was a 'vice' and not to be encouraged, since developments in that area (not to mention the spectacular increases in agricultural productivity during the twentieth century) would probably have made him feel better.

But the afterlife of Malthus's name parallels that of Francis Galton, and for very similar reasons. Malthus became associated with a far more pitiless and cruel doctrine than he ever proposed, once the 'neo-Malthusian' movement later began using his arguments in order to agitate for procedures of negative eugenics – the forced sterilisation of the poor and 'unfit'. Today, a misanthropic neo-Malthusianism informs the 'deep green' wing of ecological thinking that portrays the planet as already overpopulated by an infestation of humans. In response to a 2015 article about global warming by Elizabeth Kolbert in the *New Yorker*, for example, one reader wrote in to say: 'The world's most difficult problems would be greatly ameliorated if we were able to reduce population, and this notion should be discussed.'[37] By what means he thought that population ought to be 'reduced' was, tellingly, left unsaid. Yet what Malthus himself actually wrote is very far from the pessimistic misanthropy that is even today, usually unthinkingly, ascribed to him.[38]

In the mean time, Malthus had an important influence on the development of economics and demography. And we even have Malthus to thank for inspiring Charles Darwin and Alfred Russell Wallace in their co-discovery of the principle of evolution by natural selection. As Darwin relates in his autobiography: 'Fifteen months after I had begun my systematic enquiry, I happened to read for amusement Malthus on Population, and being well prepared to appreciate the struggle for existence which everywhere goes on from long-continued observation of the habits of animals and plants, it at once struck me that under these circumstances favorable variations would tend to be preserved and unfavorable ones be destroyed. The result of this would be a new species. Here, then, I had at last got a theory by which to work.'[39]

Will it ever be acceptable or even virtuous again to argue for limiting the right to human reproduction? China's one-child policy, as is well known, had some horrific consequences – among them, the killing of female infants by parents who wanted their one child to be a boy. And yet the one-child policy also enabled a burgeoning population to survive on its limited freshwater resources, to the extent that one environmental scientist argues that it 'saved the world'.[40] China discontinued the policy at the end of 2015, but not in favour of a Catholic-style free-for-all: the one-child limit is now a two-child limit.[41] In the mean time, the philosopher Sarah Conly has already argued, in her 2015 book *One Child: Do We Have a Right to More?*, that the answer to the question posed by her subtitle is 'No'. Having more than one child, she suggests, puts unacceptable pressure on future generations in a world of limited resources, and we ought not to impose in this way on those who come after us. Governments, therefore, could be justified in limiting the reproduction of individuals, for the benefit of all society. This could be accomplished, she suggests, through education, freely available contraception, and tax breaks for smaller families. There is no inevitable future of forced terminations and sterilisations. 'Any kind of one-child policy will be unattractive,' she writes, 'but the alternative looks to be worse.'[42]

Whether in some even less happy future we go back to Malthus or to Francis Crick, it seems that the idea that anyone can have children (if they are able to), and as many as they please, could one day seem generally wrong and socially harmful. (It could easily be considered wrong and socially harmful, for example, on a spaceship.) The question of who gets to decide who is fit to have children would, of course, always be a political quagmire, as the participants of the Ciba symposium already knew. But it's an issue that could well crop up again if resources dwindle.

The view from tomorrow

If some old ideas are to return, they will need to be shorn of their pariah status by having their evil histories put to one side. It always helps, to begin with, to scrutinise what their progenitors actually said, rather than assuming their pariah reputations are deserved. Giving a fair hearing to people from the past who have been unfairly maligned is an ethical demand of rethinking.

Talking about these ideas raises provocative questions, but no such speculation ought to be unspeakable. As Crick said: 'This isn't so wild that we need not discuss it.' Declaring anything unthinkable is an insult to all thought. It helps, perhaps, to draw off the heat of such arguments by ascribing the opinions we are considering to enlightened future generations. In general it's a useful exercise to adopt what I'll call the View from Tomorrow. It is another kind of rethinking: projecting one's mind into a more civilised future and scrutinising the present with sympathy but also with cold justice, in order to question its governing assumptions. How many other practices that today we consider completely normal will one day send a shudder down the spines of our descendants?

13

Don't Start Believing

What else might we be wrong about today? And how should we rethink our ideas about ideas?

'The man who has fed the chicken every day throughout its life at last wrings its neck instead, showing that more refined views as to the uniformity of Nature would have been useful to the chicken.'

– Bertrand Russell

The happy Sceptic

Once upon a time there was an unsuccessful painter who went travelling far and wide, and ended up founding one of the most potent traditions in all of philosophy. His name was Pyrrho, and he was born, in around 360 BCE, in the ancient Greek state of Elis, home to the Olympic Games. And he still has much to teach us.

Pyrrho began his public life by contributing what that wonderful biographer of ancient philosophers, Diogenes Laertius, calls some 'indifferent' paintings of torch races to the gymnasium in the city of Elis.[1] Then he abandoned painting to study philosophy in a Socratic school. Afterwards, Pyrrho became friends with the philosopher Anaxarchus, who was a follower of the ideas of the laughing atomist, Democritus. Anaxarchus, as it happens, was also friends with Alexander the Great. So off went Pyrrho and Anaxarchus on Alexander's campaign in India.

Not much is known about Anaxarchus's philosophy, though what is reported is highly suggestive. He likened the things we see to painted scenery, and claimed that our everyday experiences are just like those of dreamers and madmen. (In this way he arguably anticipated modern theories of perception as mental construction, and the spectrum of psychosis.) Anaxarchus was also fond of taking the great Alexander down a peg or two. When Alexander claimed that he was descended from Zeus, his friend pointed to the general's bleeding wound. 'See? The blood of a mortal!' And when Anaxarchus said that there were infinitely many worlds in the universe, Alexander despaired, seeing as he had not even conquered one yet. Evidently he was too invested in glory and success. For Anaxarchus, the wise man should be indifferent to the value of things. It seemed to work for him: he came to be called the Happy Man.[2]

So he was, apparently, even until the end. Once while dining with Alexander, Anaxarchus had insulted the Cypriot king Nicocreon, who was also present, having submitted to Alexander's rule. After Alexander died, Anaxarchus wound up in a ship that made an unexpected landing in Cyprus. Nicocreon captured him, had him put in a mortar, and ordered him to be pounded to death with iron pestles. Anaxarchus taunted him: 'You may pound the pouch containing Anaxarchus; you do not pound Anaxarchus himself.' And so he left this world.[3]

This lesson of heroic equanimity certainly seems to have rubbed off on his young friend Pyrrho – who was also influenced by conversations on his travels with Indian gymnosophists (naked wise men) and magi. What he brought back to Greece as the seed of his new thought was the joyful philosophy that became known as Scepticism.

We think of scepticism now as a rather bitter, negative thing, even if it is sometimes required of us rationally. Yet for the ancient Sceptics, beginning with Pyrrho, it was the only reliable route to true happiness and contentment. Diogenes calls it a 'noble' philosophy, based on 'agnosticism and suspension of judgement'. The guiding principle of Scepticism, indeed, was to withhold judgement about any positive proposition if total certainty could not be achieved. That attitude is what they called *epoché*. It means 'suspension': not the suspension of disbelief, as we say – but the suspension of *belief* itself. (Two millennia later, the German philosopher Edmund Husserl revived the idea: for him, suspending firm opinions about the external world enables us to examine the contents of our own consciousness without prejudice and thus achieve a greater understanding of our experience.) More broadly, *epoché* can be used to mean a suspension of judgement about any idea, a refusal to label it as correct or incorrect, or morally wrong or right. People also talk of 'bracketing' an idea so that we can use it in our subsequent thinking – held gingerly, as it were, between our fingertips – without committing to an opinion about it. (This recalls, too, the way that Stoic-influenced CBT teaches the subject to observe her own thoughts neutrally rather than being judgmental about them.)

The word *skeptikos* itself means someone who is inquiring or searching, rather than simply rejecting things.[4] It just so happens that, on his positive search for reliable knowledge, the Sceptic has not yet found any. We can always be fooled by appearances. Only by suspending judgement and refusing to enslave himself to belief, therefore, will the wise man achieve the state of *ataraxia*, a lucid tranquillity of mind. (This sounds somewhat like the Buddhist tranquillity achieved by freeing oneself from attachment, so perhaps the Indian philosophers Pyrrho met really did have an influence.)

Of course you need *some* working beliefs to survive – say, the belief that stepping off a cliff will kill you, or that some things are good to eat and others are poisonous. Later followers of Pyrrho would say that the wise man lets himself be guided by the appearances in such cases – eating an apple rather than a wasps' nest – without really believing in them.[5] Diogenes says that Pyrrho himself took no precautions at all, wandering around heedless of impending peril from wagons or precipices, and only his devoted students kept him from harm. On the other hand, he was once scared by a dog that charged at him. Someone accused him of hypocrisy, and Pyrrho replied gently that it's very hard to strip oneself entirely of human weakness. Sometimes, indeed, animals were the best models for the desired attitude of total insouciance. Once when Pyrrho was aboard a ship his fellow passengers were frightened by a storm. He kept calm, pointing to a small pig on the ship that just carried on eating its dinner. That, he pronounced, was the state to which the wise man should aspire.[6]

It may be surprising that, for someone who espoused such an apparently nihilistic philosophy, Pyrrho was widely loved. He lived with his sister, who was a midwife, and did his fair share of the dusting; he could often be seen taking pigs and chickens to the market. Pyrrho loved the work of his predecessor Democritus, and also liked to quote his Homer, often repeating the line: 'As leaves on trees, such is the life of man.' He was so widely admired by his fellow Greeks, indeed, that a law was passed making all philosophers exempt from tax.[7] Would that such a paragon lived in our time.

Pyrrho never wrote anything himself, but his followers did, and even Plato's Academy was taken over for a time by professors of Scepticism. The definitive account of this philosophy in antiquity was set down nearly half a millennium after Pyrrho's life, in the second century CE. That was done by a doctor, Sextus Empiricus, who proudly called himself a Pyrrhonist. And then the tradition more or less vanished – until its rediscovery in the Renaissance, starting a chain reaction that inspired the brilliance of the French essayist Michel de Montaigne and the British philosopher David Hume, among others. To this day, its core idea – the suspension of belief – turns out to be a critical guiding principle for both modern scientists and modern spymasters.

Soul dust

Much impressed by the manner in which Pyrrho's mentor, Anaxarchus, met his torturous death, Diogenes Laertius wrote the lines: 'Pound, Nicocreon, as hard as you like: it is but a pouch. Pound on; Anaxarchus's self long since is housed with Zeus.'[8] Anaxarchus's immortal soul, in other words, was immune from physical blows and had ascended to heaven.

That we have souls, of course, is still a belief sincerely shared by many people of a religious persuasion. According to the Pew Research Center's 2014 poll, more than two-thirds of Americans believe in an afterlife.[9] But in the world of secular philosophy and science, the soul has long been assumed to be the deadest idea possible. No one thinks our bodies are imbued with some invisible, immaterial soul-substance that animates them and hosts our thoughts and dreams. We are all hard-headed materialists; we know that physical matter is all there is (even, you will recall, if we are panpsychists). We are definitely not 'substance dualists', like Descartes, dividing the world into physical substance and mental substance.

The death of the soul may seem like a loss of spiritual consolation. But the modern analytic philosopher Peter Unger has taken the idea of the soul very seriously, and found it profoundly unconsoling. If we have immaterial souls, he asks, what will become of us? His answers do not offer much succour. Separated from a brain and physical sensory apparatus such as eyes and ears, Unger argues, an immaterial soul is likely not only to be unable to see and hear, or to experience touch, but may suffer from the dreadful condition known in physical human beings as 'prospective amnesia', in which the sufferer is unable to lay down any new memories about present experience. How, Unger asks, would you be able to lay down memories if you didn't have the electrochemical structure of neurons by which this, in actual people, is done?

'Going by our own current desires and values,' Unger writes,

> how worthwhile an existence would there be for a disembodied soul, utterly alone and experiencing always so terribly drably? Even if such an existence should not be so bad as ceaseless severe torture, it would be worse than simply ceasing to exist. This is especially clear if these experiential futures should last for billions of years, perhaps even forever, *with no way out*. [. . .] Unless almost all available evidence is badly misleading, what will become of us is nothing that any of us really wants and, beyond that, it will be something – whether of one sort or whether of another – that many of us may well dread.[10]

Gulp. But there is hope, in that Unger may well, after all, be wrong. Which he is more than happy to admit. Indeed, perhaps what is most interesting about Unger's treatment of the topic is the unabashed distinction he draws between two attitudes to an idea. 'Well, in this big neighborhood, that's what I've *argued*, and I *would still argue*,' he writes about his version of ensoulment. 'Though I've been arguing all that, I have little idea, really, about what to *believe* on these large metaphysical matters. And, so, I'm agnostic, regarding them all.'[11]

That sounds odd. In everyday life, we might be tempted to say that someone who *argues for* an idea without *believing* it is being disingenuous if not hypocritical. But Unger's distinction is sensible and even humane. Sometimes, posing a demand for belief puts too much stress on an idea, which might regardless of its truth have other possible virtues, such as fertility, a placebo effect, or the sheer interest of mapping an unexplored part of the space of possibilities. Secondly, belief is hubristic: it denies the possibility of error. And thirdly, beliefs are often very dangerous things. People die for them; and more die unwillingly for the beliefs of others. It is sometimes lamented that people in advanced countries don't believe anything any more. To the extent that this is true, it is arguably progress.

Posing the question of belief in a particular view will often bring in, too, extra-intellectual issues of social belonging or alienation. This is a possibility that Unger himself brings up later: 'I hope that my reluctance to embrace Entity Dualism, or Substance Dualism, is not due, in any great measure, [to] the fact that the View is so terribly unpopular, even so widely and deeply disdained, among almost all those in my professional culture-circle.'[12]

Of course, the fact that a view is disdained is by itself no reason to rethink it and become its champion. But neither is it any reason, by itself, to discard it. After all, reality has repeatedly shown that it doesn't care much for our disdain one way or the other.

In choosing to *argue* a particular position about immortal souls but not to *believe* it, Peter Unger is channelling Pyrrho and the Sceptical tradition. Well, you might say, a commitment to arguing but not believing may seem like a nice method in the relatively low-stakes arena of philosophical metaphysics. But we can't behave that way in the real world. Except that something very like a classical attitude of *epoché* is just what is recommended, too, in a legendary CIA manual of spycraft.

Richards J. Heuer is a CIA analyst who originally studied philosophy at university, where he became interested in the problem of epistemology, or how we can come to know things. This is an eternal theoretical problem (the Sceptics thought we can never come to know things for sure), but it is obviously also an important practical problem in the world of spies. Heuer eventually produced a widely admired CIA manual on the topic, *Psychology of Intelligence Analysis*. In it, he writes that 'the principle of deferred judgment' is the most important aspect of creative thinking. 'The idea-generation phase of analysis should be separated from the idea-evaluation phase, with evaluation deferred until all possible ideas have been brought out,' he explains. 'This approach runs contrary to the normal procedure of thinking of ideas and evaluating them concurrently.' In other words, hold all your ideas in suspension until it is time to choose one of them. Why? Because 'a judgmental attitude dampens the imagination'.[13] And a spy's imagination has to encompass vertiginous possibilities of double- and triple-bluffing. Indeed, as the former counter-intelligence chief James Jesus Angleton pointed out, the spy is, in a way, in an even more difficult position than the scientist. A biologist looking through a microscope can assume that what she sees has not been deliberately arranged there in order to mislead her. But the spy must always take into account the possibility that the enemy has conspired to produce disinformation.[14]

Committing to one interpretation too quickly, then, is dangerous for the spy. It can also be bad for business. Recall Warren Buffett's success with value investing. That strategy doesn't stoop to instant reaction on the day-to-day fluctuations of stocks. It buys and holds, suspending judgement during blips and passing crises, just as a physicist holds off discarding an idea in case it might turn out to be fertile or truthifiable. Buffett's strategy rethinks what an investor's rhythm of judgement should be. It plays the long game.

What if we thought of ideas in an analogous way? Perhaps it's also worth avoiding knee-jerk reactions when considering whether to buy or sell stock in an idea. In our modern liberal societies it is considered wrong to be too 'judgmental' about people and their lifestyles. But many people are happy to be judgmental about ideas. It is common to dismiss ideas quickly as stupid or vicious, as obviously wrong or trivial. But, as Heuer argues, that dampens the imagination. One way to liberate it is to adopt the View from Tomorrow, as we saw in the last chapter, to try to consider an idea free of the moral weight that attaches to it in particular historical circumstances. And we could try to get into the habit of deferring judgements about ideas more generally.

Employing the attitude of *epoché*, or suspension, is a critical tool of investigation, but it doesn't imply that anything goes. Recall how the wise Sceptic in Pyrrho's tradition would 'give in to appearances' in order to eat non-poisonous objects and avoid walking off cliffs. That might seem a somehow unrespectable fudge – yet it is often what modern science does. Being guided by appearances while suspending belief in their veracity is very close to the attitude in quantum physics that just concentrates on the mathematics without getting too worried about what it implies about the true nature of reality. (Such a neo-Sceptical stance is most famously summed up in the slogan: 'Shut up and calculate!', usually – but possibly erroneously – attributed to Richard Feynman.)[15]

More generally, it seems that only by suspending judgement in this way can we realise that the truth or falsity of an idea is not always the most interesting thing about it (it could be a placebo, or a stepping stone), or that an idea's apparently obvious moral value might be historically contingent (it could be a future pariah). And only by using *epoché* have the thinkers in this book noticed that a discarded idea really did have something to it after all.

Of course, at some point we will need to judge an idea. But *epoché* still has a critical place in the process. Think, for example, of what a criminal jury does. Jurors are instructed that they must not find the accused guilty unless they are convinced *beyond a reasonable doubt* – or, in the parlance of modern English law, 'sure so that you are certain'. In other words, they must suspend their judgement unless and until the case is proven.

On a more day-to-day level, sometimes we want to generate a lot of ideas and then choose the best one to pursue. Even then, *epoché* is an important part of the process. 'One trick that works for us,' write the *Freakonomics* authors Steven Levitt and Stephen Dubner, 'is a cooling-off period. Ideas nearly always seem brilliant when they're hatched, so we never act on a new idea for at least twenty-four hours.'[16]

Note that we do not use the term 'scepticism' today to mean quite the same as *epoché* meant for the Sceptics. Insofar as it is resistance to believing in any old rubbish, scepticism is a virtue, and indeed crucial to science ('Go on, impress me'); but often it will harden into polemical incredulity towards any novel or uncomfortable suggestion at all. In our age, the term 'sceptic' is often hijacked by people who flatly reject some proposition while credulously grasping at any scrap of dubious factoid or argument that seems to contradict it. This is the case, for example, with 'climate-change sceptics' who reject the very well-grounded idea of human-caused global warming. They are sometimes called 'denialists', but it is more accurate to say that their problem is a lack of *epoché*: they rush to judge the consensus idea as wrong, before they have amassed any compelling arguments for doing so. Successful science, on the other hand, has *epoché* built into it, at least in its ideal form: its top results should still always be considered provisional, approximate – good enough to work with until something better comes along. This does not mean that our current best theories are not sufficient to send robots to Mars, or to save the lives of millions of children through immunisation.

Where's the evidence?

The stance of *epoché* takes a quizzical, arm's-length approach to evidence. It also has a mirror truth, which is summed up in the familiar saying: *Absence of evidence is not evidence of absence*. Now, this phrase tends to be rolled out a lot by people who feel disenfranchised by modern science's inability to countenance their sincere beliefs in ghosts or hypersensitivity to Wi-Fi, and so most rational admirers of science's success tend to get exasperated with it. They tend to treat it as implying that anything goes. The philosopher Bertrand Russell, in reply to the objection that one cannot prove God does *not* exist, pointed out that one cannot prove that there is not a teapot currently in orbit around Mars – but that doesn't mean we have the slightest reason to believe there is one.

Yet 'absence of evidence is not evidence of absence' is a deep and powerful truth. In English and American law, a criminal jury acquits by finding the defendant *not guilty*, rather than 'innocent'. Because the absence of sufficient evidence of guilt is not the same thing as evidence of the total absence of guilt. And the principle is vividly present on the long historical view of ideas. For thousands of years there was no direct evidence for the existence of atoms. Indeed, it might be in the nature of revolutionary new ideas that, at the time they are first proposed, there is no evidence for them. Or at least, that there is no more evidence for them than there already is for the existing explanation. (Remember that Copernicus's model of the solar system did not explain the movement of the planets any more accurately than Tycho Brahe's.) True, if you do careful experiments and find nothing, you will be more inclined to think that the idea won't fly. And things can be ruled out: if theory predicts that a particle will be found within a certain energy range, and it is not found after repeated efforts, then the theory is very likely wrong. Yet this kind of precise negation of an idea is seldom achievable even in science.

It's tempting, even, to say that there is no such thing as *evidence for an idea*. On the one hand there is a stock of evidence – which is already constructed and collated according to some theory, or more usually various theories, as in the Large Hadron Collider. 'Evidence' is not a solid, unchanging and absolutely reliable ground on which various ideas can compete. Indeed, as the biologist Stuart Firestein points out: '[A]ll scientists know that it is facts that are unreliable. No datum is safe from the next generation of scientists with the next generation of tools. The known is never safe; it is never quite sufficient.'[17] On the other hand there are competing ideas, and it is surprisingly easy to corral one arbitrary subset of the existing evidence in support of any arbitrary idea you like. This is not scepticism about the reliability of science; it is realism about how science has actually been done throughout its history. (This does not mean that we should not strive for 'evidence-based' policy in public health and other arenas. But we should look at who has collected the evidence and on whose instructions – with what kind of implicit bias – and whether there is a class of data that might have counted as evidence but has been ignored.)

The parallel in this regard between scientific inquiry and the work of spies is made clear by Richards Heuer, who points out: 'No matter how much information is consistent with a given hypothesis, one cannot prove that hypothesis is true, because the same information may also be consistent with one or more other hypotheses [. . .] a long list of evidence that is consistent with almost any reasonable hypothesis can be easily made.'[18]

So it seems we'd better suspend our belief in any particular reasonable hypothesis as long as possible.

The water cure

But surely we can still rule some things out as nonsense, right? Take homeopathy. We know there are no unknowns there. The preparation of a homeopathic treatment consists of dissolving a herbal extract in water and performing 'succussion' on it (basically, banging). Then there is another round of dilution with water and succussion, and another, and another . . . until finally the liquid is so dilute that it is extremely unlikely that even a single atom of the active ingredient will be present. In that case, whatever clinical efficacy homeopathy has must be attributable to the placebo effect and nothing else. Of course, as we know, the placebo effect is mighty and mysterious. But in claiming something more than that, homeopaths are agents of nonsensical woo-woo and are – like any 'alternative' practitioner who discourages people with grave illnesses from pursuing actual medical treatment – sometimes positively dangerous.

Yes. And yet – as Michael Brooks, a science writer with a doctorate in physics, points out – water itself is still surprisingly little understood. 'We know very little about liquids,' Brooks says, and 'water is a particularly strange liquid.'[19] Water's notorious weirdness – including the fact that ice is less dense than liquid water, and floats on it – is probably the reason there can be life on Earth in the first place. Water molecules can cluster in unexpected geometrical forms (for instance, in the regular twenty-sided shape called an icosahedron); they can form 'beads' and 'chains'. A quantity of water, at the microscopic level, is not at all homogeneous in the way we assume. Even a molecular-level understanding of how water evaporates remained elusive until very recently.[20] And if you grant the possibility in water of epitaxy – 'a well-known phenomenon,' Brooks explains, 'in which structural information is transferred from one material to another' – then you have a speculative mechanism whereby the homeopaths' claims that water can retain a 'memory' of what was once dissolved in it could actually work. (It is the molecular *structure* of substances, and nothing else, that gives them their everyday qualities.)[21]

More probably, though, homeopathy would work much better if the huge dilutions didn't happen at all. Brooks describes some intriguing research in mathematically enhanced cladistics (the classification of biological organisms) that suggests the remedies which homeopaths group together to treat particular bodily systems actually share chemical similarities. 'Which means,' Brooks writes, 'that dilution and succussion [. . .] could be not just a waste of time but the root of homeopathy's problems. If its power lies in chemistry, there is no need to jump through the hoops of imprinting structure on liquids.'[22] Extracting proven remedies from plants is, after all, perfectly respectable in medicine. Aspirin (from the bark of the willow tree) is the headline example, but Big Pharma does it all the time. And the Chinese researcher Tu Youyou got her Nobel Prize for finding a way to exploit the potency of sweet wormwood to combat malaria.

In the nineteenth century the British doctor Richard Hughes became editor of the *Annals of the British Homeopathic Society*, and he caused a fuss by calling for much less dilution of the plant extracts involved; his reward was hostility from other practitioners and the effective erasing of his name from the official history of homeopathy after his death.[23] So maybe the real problem with homeopathy all this time has been that its adherents have defended an orthodoxy that needlessly limits their own therapy's effectiveness. Sociologically speaking, the trouble for homeopaths, if that turns out to be true, is that without dilution and succussion homeopathy is no longer a separate 'idea' of its own at all. All the elaborate ritual and rationalisation is revealed as empty, and what used to be 'homeopathy' just falls back in to the long and noble tradition of seeking out the medicinal properties of natural substances.

Homeopathy might, after all, have been a black-box idea – one that worked (a bit) even though no one could figure out why – until mathematical cladistics revealed a pattern. The more general problem that example illustrates is that, because of the way science demands precise causal explanations – as it must – it can often get black boxes wrong. When someone proposes a phenomenon for which no one as yet understands the mechanism, like the inheritance of acquired characteristics or the reduction of patient death through handwashing, science will often reject it. Partly because science – like Max Planck's teacher – tends to assume that it knows pretty much everything about what kinds of causes and forces exist. And those assumptions are what enable it to make useful predictions, so that our aeroplanes and computers work the way we expect.

A black box in an aeroplane contains specific information to solve a puzzle; by contrast, a black-box idea has something irreducibly mysterious at its core, something that must be taken on faith. That is why sorting the good black-box ideas from the bad ones presents such a challenge. To be fair, the scientists who rejected Lamarckism with such force throughout the twentieth century thought they had hard evidence against it – the old picture of genes as unalterable legacy. But this evidence suddenly seemed much weaker in the face of the discoveries of epigenetics, which provided the crucial mechanism needed for researchers to take a kind of neo-Lamarckism seriously again. And this raises questions in other fields. Might researchers one day suggest a plausible mechanism for the alleged psychic phenomena that interest Rupert Sheldrake, such as pet telepathy? It sounds fanciful. But what history tells us, time and again, is that absolute confidence is a shaky position at best.

Cut off the bias

In 1998, the Indian government tested a nuclear weapon, which came as a surprise to US analysts and therefore became a standard example of an 'intelligence failure' on their part. Richards Heuer says that this failure can be ascribed to the overvaluing of Strategic Assumptions and the undervaluing of countervailing Tactical Indicators. In the Indian case, the Strategic Assumption was 'that the new Indian Government would be dissuaded from testing nuclear weapons for fear of US economic sanctions', while the Tactical Indicators available at the time included reports that the Indian government was actually preparing to test a nuclear weapon.[24] Amazing as it may seem, the intelligence community decided that those reports must be false, since they believed so strongly that it couldn't happen. Similarly, Aristotle rejected the atomic hypothesis because his Strategic Assumption (that the four elements were indivisible) was immune to the Tactical Indicators of Democritus's concrete arguments about the smell of bread and the evaporation of puddles. Aristotle – such a brilliant thinker in many other ways – did not want to suspend that assumption in order to allow new arguments to change his mind. But *epoché* teaches us to suspend belief even in the most apparently robust Strategic Assumptions, to avoid being caught out.

The news is even better than that. To adopt the Sceptics' attitude of *epoché* is not just to avoid being taken in by some idea that looks persuasive but is flawed – it frees us to be the best thinkers we can be. We've seen that there can be very good reasons to reject promising ideas. (Tycho Brahe's attack on Copernicanism was in many ways correct.) But there are also bad reasons. And a lot of them are driven by our cognitive biases – the mental blips that Jochen Runde talked about when discussing business strategy. We are often easy prey to the availability heuristic (thinking that events which come readily to mind are more common) or the confirmation bias (taking account only of observations that confirm our view). Not all of these biases are necessarily problematic – and, as I have argued elsewhere, their existence certainly does not add up to a proof that human beings are generally 'irrational'.[25] Still, they can be traps.

Luckily, we have an excellent defence against them: *epoché* itself. Suspending belief protects us from the dangers of making a snap judgement, where a Sceptical attitude would be more constructive and fertile. Once we are aware of the dangers of cognitive biases, we can do something about them. Richards J. Heuer, too, recommends that intelligence analysts be aware of cognitive biases and suspend their judgement as long as possible while surveying competing ideas. In doing so, he writes, 'One becomes less confident of what one thought one knew.'[26] And that reduction of confidence opens a space for new perspectives. In the battle against our own cognitive biases, the ability to suspend our belief may well be the best weapon we have.

Throughout this book we have observed many ways of thinking about ideas, both old and new. And they all depend on maintaining the suspension of belief. That gives us time to look at an idea from other, potentially more fruitful angles, and perhaps to rediscover its continuing power. Take an idea and see if it might be a black box (Lamarckism). Forget about whether an idea is true or false, and consider its possible placebo effect (William James's theory of emotion), or whether it is a necessary stepping stone even if it might be wrong (dark energy). Ask whether an idea has been rejected not because it is stupid but because it would be a power-up (multilingual computer programming). Pay serious attention to the least ridiculous option (panpsychism). Identify what we know we *don't* know, to stimulate curiosity. Abandon common sense and bet against the market. Take another look at what seems too simple to work. Adopt the View from Tomorrow for an enlightening perspective on current thinking. The suspension of belief is a powerful engine of discovery, and of rediscovery.

14

Epilogue: Back to the Future

'The time will come when mental acumen and prolonged study will bring to light what is now hidden . . . the time will come when our successors will wonder how we could have been ignorant of things so obvious.' – Seneca

Our culture is glutted with narratives of bygone wars and dramatisations of vanished societies, yet it pays far less attention to history's great struggles of thought. And any culture that thinks the intellectual past is irrelevant is one in which future invention threatens to stall. The most formidable innovators are those who know their history, like Elon Musk and Isabelle Mansuy. We need CRISPR and Twitter, but also Grace Hopper and Pyrrho. This book's highly selective snapshot of the looping evolution of ideas, after all, implies very strongly that even after everything from e-cigarettes to epigenetics, placebo ideas, and Social Dividend, new approaches to many other problems might still lie slumbering in the past, waiting for their own moment of rethinking and rediscovery.

'Those who cannot remember the past are condemned to repeat it,' said George Santayana – and the same goes for the history of ideas. Virtually none of the debating points in the great new-atheism struggles of the twenty-first century, for instance, would have been unfamiliar to medieval monks, who by and large conducted the argument on a more sophisticated and humane level. Ignorant replication gets us nowhere. But a powerful way to 'Think Different', in Apple Computer's former slogan, is to consciously rethink something old, like the electric car or Lamarckian inheritance. Society worships originality, but originality is overrated. (If I have accidentally had any new ideas in this book, you should probably take them with a pinch of salt.) No one should be ashamed if it is pointed out that someone has had their idea before. Rejoice! Perhaps it can be made even better.

Call no man happy until he is dead, said the ancient Athenian statesman Solon. After all these stories of the return of ridiculed ideas, we might vary the formula to say: call nobody wrong until the end of time. The present day is not the End of History in ideas any more than it was in geopolitics when Francis Fukuyama first announced it in 1989. We are simply living through another historical stage of error and evolution. Truth, like Paris, may be a moveable feast. But that doesn't have to stop us searching for it, in the past and in the future.

The final frontier

The Space Shuttle *Columbia* disintegrated thirty-five miles above the Earth's surface on 1 February 2003. A loose piece of heat-shield insulation had gouged a hole in its wing, allowing hot re-entry gases into the craft. The shuttle's break-up killed everyone on board, and also killed the space age. Or so it seemed at the time. NASA cancelled plans for manned missions beyond the International Space Station. No Shuttle flew for two and a half years, and the launch schedule slowed, until the programme was eventually cancelled. The last Shuttle mission ever to fly was *Atlantis* in 2011. With war, famine, and global warming right here on planet Earth, a view had already become increasingly mainstream over the previous decade that space exploration was too whimsical and expensive a distraction. It seemed obvious that we ought to sort out our human problems at home before spending billions flying around the inky void in the hope of finding something interesting. We still sent out probes and landers, and got a remote-controlled robot to Mars, but sending humans beyond low-Earth orbit seemed like improbable science fiction once again. If, on occasion, you looked up at the Moon and reminded yourself that human beings had actually flown there and walked about on it, the whole thing seemed amazing, a Homeric feat of ancient heroes that we could no longer hope to replicate.

Space exploration has been a paradigmatic vision of the future for well over a century. Yet every actual space mission is an old idea, and not just because the planning can take a decade or more. The basic principle of modern rocketry is contained in the first known steam-powered machine, the aeolipile, invented by Hero of Alexandria in the first century CE. (A sphere containing water is heated, and steam jets out of two funnels on opposite sides, causing it to rotate.) And writers imagined manned voyages to space, as well as machines travelling further on their own, long before it was possible. The Greek satirist Lucian of Samosata wrote of a journey to the Moon in the second century CE. (The Moon, it turned out, was populated by three-headed vultures and creatures who were half-woman, half-grapevine.) In the modern age, popular science fiction both influences and reflects ideas in contemporary science. There is now real physics on the subject of time machines, and the cosmologist Lawrence M. Krauss's celebrated book *The Physics of Star Trek* ushered in a growing genre that examines the concrete possibilities of teleportation or faster-than-light warp drives. But towards the end of the Space Shuttle's career, when it had begun to seem more of a glorified Staten Island Ferry to low-Earth orbit than a forerunner of the USS *Enterprise*, popular science fiction turned inward-looking, virtual, or philosophical. The model was *The Matrix* and *Inception* rather than *2001: A Space Odyssey*.

Space travel is inherently dangerous. In the hostile vacuum of space, you have to take your own atmosphere and environment with you, wrapped in thin metal. And yet some people still thought it was worth it. A clique of writer enthusiasts had kept the faith, producing the kind of near-future hard science fiction in which humanity kept exploring the stars. Sometimes it was grounded in an alternative NASA history of the past few decades. The novelist Stephen Baxter did both. Baxter, trained in aerospace engineering, has named H. G. Wells as a huge influence on his work; indeed, he wrote an 'authorized' sequel to *The*

Time Machine. (Even SF, the most forward-looking branch of literature, reaches back to old ideas, especially when reality is failing to provide any new inspiration.) And some of Baxter's work looked eerily prescient: in a jolting, terrifying sequence in his 1997 novel *Titan*, a Space Shuttle breaks up on re-entry into the atmosphere, though the crew manages to bail out and survive. The name of that shuttle? *Columbia*.

And then something unexpected happened – everyone seemed to get excited about space again. In August 2014, the European Space Agency's craft *Rosetta* successfully intercepted the snappily named comet 67P/Churyumov-Gerasimenko. *Rosetta*, a 3,000-kilogram craft with solar-panel wings spanning thirty-two metres, had been chasing the comet's tail for ten years, employing four separate gravity assists by slingshotting around Mars (once) and Earth (three times) in order to match speed and direction with its target. Then, on 12 November, *Rosetta* launched its lander *Philae* on to the comet's surface. The plucky lander, the size of a washing machine, bounced twice, sending back photographs, and then came to rest in the shadow of a crater wall, while watchers on social media held their breath. Without enough sunlight to its solar panels, its battery would be depleted in sixty hours. In the event, *Philae* managed to send back 80 per cent of the mission's planned surface data measurements before going off-air just after midnight on 15 November. One of the most important things it discovered was that complex organic molecules such as acetone and formaldehyde were present, implying that the 'building blocks of life' were present early in the formation of the Solar System.[1]

Undiscouraged by its lander's difficulties, the mother ship *Rosetta* continued to escort the comet in its orbit around the Sun into 2015, analysing the streams of gas it regularly gave off: it turned out that they contained oxygen molecules, showing that oxygen too must have been present when the comet formed.[2] And, thrillingly for anthropomorphic robot-fanciers, *Philae* eventually woke up again. Harvesting minutes of sunlight per day from its awkward position on the comet, it had slowly recharged its battery sufficiently to come out of safe mode. Unfortunately it was only a brief resurrection, and *Philae* was not heard from again after 9 July. But another spacecraft was soon taking over the public's attention. NASA's 500-kilogram *New Horizons* probe, the size of a piano, had been hurtling through the solar system for nine years, at an average speed of more than 30,000 miles per hour. On 14 July it arrived at Pluto and its moons, and sent back astonishing close-up images of the tiny planet for the first time. Most people had assumed Pluto was just a dead, battered snowball; in fact it had towering ice-capped mountains and glacial activity that was slowly carving valleys into its surface, indicating an unexpected internal heat source at the planet's core.[3] It also turned out that the planet had a surprisingly extensive atmospheric 'haze' of soot-like particles, giving it a blue sky in the daytime.[4]

In the age of Twitter and global enthusiasm for science, these were major cultural events. (Vangelis had been commissioned to compose music for the *Rosetta* mission's promotional videos, and XKCD comic author Randall Munroe drew a live updating comic on his website the day of *Philae*'s landing.) But perhaps the most important event, at least for what it implied about the future of space travel, was the successful landing of another craft back on the Earth's surface.

Not content with reinventing the electric car, Elon Musk also has a commercial space company, SpaceX, which has contracts with Nasa to send up supplies to the International Space Station, and with private companies to put satellites into orbit. (Another example of science fiction inspiring reality: its rocket is called the Falcon, in homage to the Millennium Falcon in *Star Wars*.) SpaceX's rocket factory in Los Angeles defies aerospace-industry convention by building nearly all its components – from engines to electronics – itself, which enables it to modify and iterate its designs very quickly. (Insourcing is the new outsourcing.)[5] SpaceX's rockets had suffered some teething problems – one on an ISS resupply mission exploded in the summer of 2015, suffering what Musk laconically described as a 'rapid unscheduled disassembly'. (Some Flat Earthers, inevitably, took this as proof that the rocket had hit the impermeable dome that covers our discworld planet.) But in December, an improved Falcon 9 rocket launched from

Cape Canaveral, and successfully deployed nine communications satellites into orbit. Then its main-stage booster, the size of a fifteen-storey building, did what no non-test rocket had ever done before. Having fired off its cargo, it made a screeching U-turn while moving at 3,000 miles an hour, burned back through the Earth's atmosphere, and landed right where it had taken off, descending gently on its own fiery exhaust and settling on deployable legs. Ten days later Musk posted a picture of the rocket on Instagram: 'Falcon 9 back in the hangar at Cape Canaveral. No damage found, ready to fire again.' So this was the first reusable orbit-capable spacecraft since the Shuttle. And being able to reuse your hardware makes spaceflight a lot more cost-effective. How much more cost-effective? 'It would be the difference between something costing less than one percent of what it would be otherwise,' Musk explained. 'It makes all the difference in the world. It's absolutely fundamental. And I think it really quite dramatically improves my confidence that a city on Mars is possible. You know, that's what all this is about.'[6]

Wait, a *city on Mars*? Yes, suddenly old space-age dreams were alive and well again, and funded by determined billionaires. Once again, a newly attractive glamour attached to the kind of can-do, gung-ho practicalities familiar from a previous golden age of science fiction in the second half of the twentieth century. This was reflected in the popular consciousness by films such as *Interstellar*, *Gravity*, and, in particular, Ridley Scott's *The Martian*, where science and engineering, as much as Matt Damon, are the heroes. 'The goal of sending a man to Mars is so much more inspiring than what other people are trying to do in space,' the PayPal co-founder Peter Thiel told Elon Musk's biographer. 'It's this going-back-to-the-future idea. There's been this long wind-down of the space program, and people have abandoned the optimistic visions of the future that we had in the early 1970s. SpaceX shows there is a way toward bringing back that future. There's great value in what Elon is doing.'[7]

But going to Mars, for Elon Musk, isn't just about making fantasy come true. It's about saving humankind. As long as we exist on only one planet, we are vulnerable. Even if the worst consequences of global warming can be averted – which Musk himself is attempting to do with his Tesla electric cars and his solar-energy business, SolarCity – we could be wiped out by an asteroid, such as the one only ten kilometres across that killed the dinosaurs sixty-six million years ago. Having a colony on another planet is insurance: it's about the survival of the species. As SpaceX president and COO Gwynne Shotwell puts it: 'If you hate people and think human extinction is okay, then fuck it. Don't go to space. If you think it is worth humans doing some risk management and finding a second place to go live, then you should be focused on this issue and willing to spend some money.'[8]

The prospect of mass extinction caused by an asteroid strike is also the motivation behind another idea to repurpose an old technology: nuclear weapons. Nukes have so far not destroyed humanity – thanks in part, as we have seen, to the efforts of nuclear strategists – and one day they even might save it. If we detect a big space rock on a collision course with Earth, it will probably be a very good idea to send up a spacecraft with a nuclear warhead to intercept it. Not so that Bruce Willis can tunnel into the asteroid and blow it up, as in *Armageddon*. Blowing up an asteroid is a very bad idea, since many of the remaining chunks will just carry on towards Earth and rain down to cause holy hell. Instead, as Nasa explains, neutrons from a detonation just above the asteroid would irradiate an area of the surface, which would then blow off and produce a slight recoil on the rest of the asteroid, altering its velocity just enough so that it eventually misses our planet completely. 'The trick is to gently nudge the asteroid out of harm's way and not to blow it up.'[9] Russian rocket scientists want to upgrade the intercontinental ballistic missiles (ICBMs) that were pointed at the US during the cold war so that they can be used as delivery vehicles for a humanity-saving killer-asteroid strike, and have suggested that this method could be tested on the asteroid known as Apophis, which may pass worryingly near Earth in 2036.[10]

In the mean time the EU-led Neoshield-2 project, a global consortium of scientific institutions and aerospace companies, is investigating the plausibility of deflecting asteroids by crashing into them with 'a high-velocity kinetic impactor spacecraft', while the DE-STAR project on 'Directed Energy Planetary Defence', run by the Experimental Cosmology Group at the University of California Santa Barbara, is investigating ways to use space-based lasers to burn off a section of an asteroid's surface and alter its trajectory.[11] Increasingly, government and scientific bodies consider this a risk worth taking very seriously. An asteroid that could kill everyone on Earth *will* be hurtling towards us at some point: the only question is when. If it's a long time in the future, it may be easy to handle with robots and lasers. Sooner than that, and we might have cause to be very glad that humans invented nuclear weapons after all.

Another space-based idea returned to prominence in the summer of 2015 with a $100 million endowment by the Russian investor and former physicist, Yuri Milner: scanning the skies for evidence of other civilisations in space. Milner's Breakthrough Listen project turbo-charged the community around SETI, the search for alien intelligence. Why now? 'In the 1960s, Frank Drake did pioneering SETI work and it was partially funded through the 1980s,' Milner said. 'But then the idea sort of faded away – except for the question of course. Today several things have changed that will allow us in one day to pull down as much data as we're currently doing in a year. We now know for a fact that there are candidates in the galaxy, a few billion. Telescope time used to be harder to get, but now there is an opportunity for private endeavors to buy telescope time. And finally Moore's Law: we can design a backend infrastructure capable of processing huge amounts of data much faster than ever before. We want to marry the best of Silicon Valley's capabilities with the best science can offer.' Milner personally thinks we're not alone in the vastness of the universe. 'If we were alone it would be such a waste of real estate.'[12]

In 2016 NASA chose a new shuttle to supply the ISS, whose tangled history was itself a little parable of rethinking. The beautiful new spaceplane, with wings that curved upwards at the tips, was actually based closely on an old Soviet design that first flew in the 1960s, called the BOR. It was more advanced than anything the Americans had at the time. In 1982, a BOR-series spacecraft splash-landed in the Indian Ocean, where it was photographed by an overflying Australian spy plane. These photos were passed to the CIA at Langley, where engineers reconstructed the design, tested it in wind tunnels, and were hugely impressed by its aerodynamic capabilities. NASA planned to build one of its own to complement its new Space Shuttle, but never did. Eventually, though, NASA's reworked design of the old Soviet spacecraft became the direct inspiration for the engineers who built the new ship.[13] That spacecraft's name? Dream Chaser.

So it was, in the second decade of the twenty-first century, that reality started to catch up with fiction again. Russia announced plans for new manned space missions, and China was aiming to land a probe on the dark side of the Moon.[14] In late 2015, following SpaceX's lead, Nasa published a remarkable roadmap for putting people on the Red Planet. In it, the agency sounded more confident than ever: 'NASA is leading our nation and our world on a journey to Mars. Like the Apollo Program, we embark on this journey for all humanity. [. . .] We are developing the capabilities necessary to get there, land there, and live there.'[15]

On planet Earth, we are all in the gutter. But maybe it's just as well that some of us are looking at the stars.

Acknowledgements

This book owes much to the advice of Jon Elek, Harry Scoble, Daniel Loedel, and Schuyler W. Henderson. Thanks to Isabelle Mansuy, Jochen Runde, Andrew Pontzen, Galen Strawson, Rupert Sheldrake, Paul Fletcher, Barbara Jacobson, Marlies Cunnen, Iain Chambers, Carl Cederström, Tony Yates, Nigel Wilcockson, Nick Humphrey, Lucien Jones, and the staff of the British Library. And particular thanks to Izzy Mant, for all the coffee and everything else.

London, April 2016

Bibliography

Abrahams, Marc, *This Is Improbable* (London, 2014)
Adamson, Peter, *Philosophy in the Hellenistic and Roman Worlds* (Oxford, 2015)
Albats, Yevgenia, trans. Catherine A. Fitzpatrick, *KGB: State Within a State* (London, 1995)
Allan, Tony, *Virtual Water* (London, 2011)
Anderson, Curtis Darrel & Anderson, Judy, *Electric and Hybrid Cars: A History* (London, 2005)
Andrew, Christopher & Mitrokhin, Vasili, *The Mitrokhin Archive: The KGB in Europe and the West* (London, 1999)
Aubrey, John, ed. Andrew Clark, *Brief Lives: Chiefly of Contemporaries* (Oxford, 1898)
Ayres, Ian, *Super Crunchers* (London, 2007)
Bacon, Francis, trans. R. Ellis & James Spedding, *Novum Organum* (New York, 1905)
Ball, Philip, *Curiosity: How Science Became Interested in Everything* (London, 2012)
——, *The Devil's Doctor* (London, 2006)
Benedetti, Fabrizio, *Placebo Effects: Understanding the Mechanisms in Health and Disease* (Oxford, 2009)
Beyer, Kurt, *Grace Hopper and the Invention of the Information Age* (London, 2009)
Bobbitt, Malcolm, *Taxi! The Story of the London Taxicab* (London, 1998)
Bohr, Niels, *Atomic Physics and Human Knowledge* (New York, 1958)
Brent, Joseph, *Charles Sanders Peirce: A Life* (Bloomington, 1998)
Brooks, Michael, *13 Things That Don't Make Sense* (London, 2009)
——, *At the Edge of Uncertainty* (London, 2014)
Burger, Edward B. & Starbird, Michael, *The 5 Elements of Effective Thinking* (Princeton, 2012)
Burns, David D., *Feeling Good: The New Mood Therapy* (Kindle edn, 2012)
——, *The Feeling Good Handbook* (New York, 1999)
Capri, Anton Z., *Quips, Quotes, and Quanta: An Anecdotal History of Physics* (London, 2011)
Chaucer, Geoffrey, ed. Larry D. Benson, *The Riverside Chaucer* (1987; Oxford, 2008)
Chomsky, Noam, *Language and Problems of Knowledge* (Cambridge, Mass., 1988)
Coué, Emile, *Self-mastery through Conscious Autosuggestion* (1920; New York, 2007)
Crick, Francis, *Of Molecules and Men* (1966; Seattle, 1967)
Dawkins, Richard, *The Extended Phenotype* (Oxford, 1982)
Diogenes Laertius, ed. R. D. Hicks, *Lives of Eminent Philosophers* (Cambridge, Mass., 1925; 1972)
Dyson, Freeman J., *Disturbing the Universe* (New York, 1979)

BIBLIOGRAPHY

Eddington, Arthur S., ed. H. G. Callaway, *The Nature of the Physical World* (1928; Newcastle, 2014)

Edgerton, David, *The Shock of the Old: Technology and Global History since 1900* (2007; Oxford, 2011)

Ensmenger, Nathan, *The Computer Boys Take Over* (Cambridge, Mass., 2010)

Epictetus, trans. P. E. Matheson, *Discourses Books 1 & 2* (New York, 2004)

Epstein, Edward Jay, *Deception: The Invisible War between the KGB and CIA* (New York, 1989)

Feyerabend, Paul, *Against Method*, 4th edn (1988; London, 2010)

——, *Farewell to Reason* (London, 1987)

Firestein, Stuart, *Ignorance: How It Drives Science* (Oxford, 2012)

Freedman, Lawrence, *Strategy: A History* (2013; Oxford, 2015)

Freud, Sigmund, trans. Anthea Bell, *A Case of Hysteria (Dora)* (Oxford, 2013)

Galton, Francis, *Inquiries into Human Faculty and Its Development*, 2nd edn (1907; Public Domain/Kindle)

Golitsyn, Anatoliy, *New Lies for Old: The Communist Strategy of Deception and Disinformation* (London, 1984)

Gould, Stephen Jay, *Ontogeny and Phylogeny* (London, 1977)

Graeber, David, *The Utopia of Rules* (London, 2015)

Grant, Edward, *Planets, Stars, and Orbs: The Medieval Cosmos, 1200–1687* (Cambridge, 1996)

Grof, Stanislav & Halifax, Joan, *The Human Encounter with Death* (New York, 1977)

Habermas, Jürgen, *The Future of Human Nature* (Oxford, 2003)

Harris, Sam, *Free Will* (New York, 2012)

Hawthorne, John, *Metaphysical Essays* (Oxford, 2006)

Heath, Chip & Heath, Dan, *Made to Stick: Why Some Ideas Take Hold and Others Come Unstuck* (London, 2007)

Helmholtz, Hermann von, ed. James P. C. Southall, *Helmholtz's Treatise on Physiological Optics* (New York, 1925)

——, ed. David Cahan, *Science and Culture: Popular and Philosophical Essays* (Chicago, 1995)

——, trans. E. Atkinson, *Popular Lectures on Scientific Subjects, Vol. II* (1895; London, 1996)

——, trans. Alexander Ellis, *On the Sensations of Tone as a Physiological Basis for the Theory of Music* (1875; Bristol, 1998)

Heuer, Richards J., *Psychology of Intelligence Analysis* (Virginia, 1999)

Holton, Gerald, *The Scientific Imagination* (1978; Cambridge, Mass., 1998)

Hoyle, Fred, *Home is Where the Wind Blows* (Mill Valley, Calif., 1994)

——, *The Nature of the Universe* (Oxford, 1960)

James, William, ed. Giles Gunn, *Pragmatism and Other Writings* (London, 2000)

Johnson, Scott C., *Ghost in the Cell* (Matter, 2013)

Jones, Lucien, *The Transparent Head* (Cambridge, 2007)

Keen, Andrew, *The Internet Is Not the Answer* (London, 2015)

Kerr, Margee, *Scream* (New York, 2015)

Keynes, John Maynard, *The General Theory of Employment, Interest and Money* (1936; New Delhi, 2008)

Kirk, Robert G. W. & Pemberton, Neil, *Leech* (Chicago, 2013)

Kondo, Marie, trans. Cathy Hirano, *The Life-Changing Magic of Tidying* (London, 2014)
Koyré, Alexandre, *From the Closed World to the Infinite Universe* (1957; Baltimore, 1968)
Kuhn, Thomas S., *The Structure of Scientific Revolutions* (1962; Chicago, 2012)
Lederman, Leon, *The God Particle* (London, 1993)
Levitt, Steven D. & Dubner, Stephen J., *Think Like a Freak* (London, 2014)
Levmore, Saul & Nussbaum, Martha C. (eds), *The Offensive Internet* (Cambridge, Mass., 2010)
Lewis, David, *On the Plurality of Worlds* (1986; Oxford, 2001)
Lightman, Alan P., *The Discoveries: Great Breakthroughs in Twentieth-Century Science* (Toronto, 2005)
Livio, Mario, *Brilliant Blunders* (New York, 2013)
MacIntyre, Alasdair, *After Virtue: A Study in Moral Theory* (1981; London, 2011)
MacKellar, Calum & Bechtel, Christopher (eds), *The Ethics of the New Eugenics* (New York, 2014)
Malthus, Thomas Robert, *Population: The First Essay* (1798; Ann Arbor, 1959)
Marcuse, Herbert, *One-Dimensional Man* (1964; London, 1991)
Meulders, Michel, *Helmholtz: From Enlightenment to Neuroscience* (Cambridge, Mass., 2010)
Montaigne, Michel de, trans. M. A. Screech, *The Complete Essays* (London, 2003)
Morozov, Evgeny, *To Save Everything, Click Here* (2013; New York, 2014)
Mount, Ferdinand, *Full Circle: How the Classical World Came Back to Us* (2010; London, 2011)
Nagel, Thomas, *Mind and Cosmos* (Oxford, 2012)
Nietzsche, Friedrich, trans. Helen Zimmern, *Beyond Good and Evil*, in Manuel Komroff (ed.), *The Works of Friedrich Nietzsche* (New York, 1931)
Petersen, William, *Malthus* (London, 1979)
Plato, *Timaeus*, in Desmond Lee (ed.), *Plato: Timaeus and Critias* (London, 1977)
——, *Charmides*, in Robin Waterfield (ed.), *Plato: Meno and Other Dialogues* (Oxford, 2005)
Poole, Steven, *Unspeak* (London, 2007)
Quiggin, John, *Zombie Economics: How Dead Ideas Still Walk Among Us* (Princeton, 2010)
Ridley, Matt, *The Evolution of Everything* (London, 2015)
Robertson, Donald, *The Philosophy of Cognitive-Behavioural Therapy* (London, 2010)
Schelling, Thomas C., *Arms and Influence* (1966; Cambridge, Mass., 1977)
Schmidt, Eric & Rosenberg, Jonathan, *How Google Works* (London, 2014)
Schopenhauer, Arthur, trans. E. F. J. Payne, *The World as Will and Representation* (1881; New York, 1969)
Schrödinger, Erwin, *What Is Life?* (Cambridge, 1944)
Scruton, Roger, *The Uses of Pessimism* (Oxford, 2010)
Searle, John R., *Mind: A Brief Introduction* (Oxford, 2004)
Sheldrake, Rupert, *A New Science of Life* (London, 2009)
——, *The Science Delusion* (London, 2012)
Skrbina, David, *Panpsychism in the West* (Cambridge, Mass., 2005)
Smith, Andrew F., *The Oxford Companion to American Food and Drink* (Oxford, 2007)
Smith, Jon Maynard, 'The Concept of Information in Biology', in Paul Davies & Niels

Henrik Gregersen (eds), *Information and the Nature of Reality* (2010; Cambridge, 2014)
Stanton, Doug, *Horse Soldiers: The Extraordinary Story of a Band of US Soldiers Who Rode to Victory in Afghanistan* (New York, 2009)
Stiglitz, Joseph, *The Great Divide* (New York, 2015)
Strawson, Galen et al., ed. Anthony Freeman, *Consciousness and its Place in Nature* (Exeter, 2006)
Sun Tzu, trans. Lionel Giles, *The Art of War* (1910; Kindle edn)
Tett, Gillian, *The Silo Effect* (London, 2015)
Unger, Peter, *Empty Ideas* (Oxford, 2014)
Vance, Ashlee, *Elon Musk: How the Billionaire CEO of SpaceX and Tesla Is Shaping Our Future* (London, 2015)
Weinberg, Gerald M., *The Psychology of Computer Programming* (1971; New York, 1998)
Weinberg, Steven, *To Explain the World: The Discovery of Modern Science* (London, 2015)
Weismann, August, trans. J. Arthur & Margaret R. Thomson, *The Evolution Theory* (2 vols., London, 1904)
Wilczek, Frank, *A Beautiful Question: Finding Nature's Deep Design* (London, 2015)
——, *The Lightness of Being* (New York, 2009)
Wiseman, Richard, *The As If Principle* (London, 2012)
Wolstenholme, Gordon (ed.), *Man and His Future* (1963; London, 1967)
Wootton, David, *The Invention of Science* (London, 2015)

Notes

Introduction: The Age of Rediscovery

1. Anderson & Anderson, 27–30.
2. Bobbitt, 7–20.
3. Vance, 316.
4. Tad Friend, 'Plugged In', *New Yorker*, 24 August 2009.
5. Vance, 312.
6. Joe Wiesenthal, 'The Tesla Model S Just Got the Best Safety Rating of any Car in History', *Business Insider*, 20 August 2013.
7. Keen, 92.
8. Andrew Smith, 'Meet Tech Billionaire and Real-life Iron Man Elon Musk', *Telegraph*, 4 January 2014.
9. Feyerabend (1988), 116.
10. Richard Conniff, 'Alchemy May Not Have Been the Pseudoscience We All Thought It Was', *Smithsonian Magazine*, February 2014.
11. 'Recreating Alchemical and Other Ancient Recipes Shows Scientists of Old Were Quite Clever', American Chemical Society, 5 August 2015.
12. 'When Woman Is Boss', *Collier's*, 30 January 1926.
13. Helmholtz (1895), 228.
14. This passage does not occur in James's 1896 lecture 'The Will to Believe', to which it is usually credited.

2: The Shock of the Old

1. Freedman, 129–30.
2. Stanton, 116.
3. Alex Quade, 'Monument Honors US "horse soldiers" Who Invaded Afghanistan', CNN, 7 October 2011.
4. Stanton, 159.
5. Ibid., 257.
6. Quade, op. cit.
7. Dwight John Zimmerman, '21st Century Horse Soldiers – Special Operations Forces and Operation Enduring Freedom', Defense Media Network, 16 September 2011.

8. 'US Special Forces Joined Charge on Horseback against Taliban', Bloomberg, 15 November 2001; Stanton, 172.
9. Freedman, 504.
10. Quoted ibid., 188.
11. Ibid., 216.
12. Donald H. Rumsfeld, 'Transforming the Military', *Foreign Affairs*, May/June 2002.
13. Stanton, 29.
14. Freedman, 222.
15. Ibid., 237.
16. Ibid., 205.
17. Lieutenant General David H. Petraeus, 'Learning Counterinsurgency: Observations from Soldiering in Iraq', *Military Review* Jan.–Feb. 2006, 45–55.
18. John Colapinto, 'Bloodsuckers', *New Yorker*, 25 July 2005.
19. Ibid.
20. M. Derganc & F. Zdravic, 'Venous Congestion of Flaps Treated by Application of Leeches', *British Journal of Plastic Surgery*, vol. 13 (July 1960), 187–92.
21. Kirk & Pemberton, 166–8.
22. Lawrence K. Altman, 'The Doctor's World; Leeches Still Have Their Medical Uses', *New York Times*, 17 February 1981.
23. 'FDA Approves Leeches as Medical Devices', Associated Press, 28 June 2004.
24. Mohamed Shiffa et al., 'Comparative Clinical Evaluation of Leech Therapy in the Treatment of Knee Osteoarthritis', *European Journal of Integrative Medicine*, vol. 5, no. 3 (June 2013), 261–9.
25. Andreas Michalsen et al., 'Effectiveness of Leech Therapy in Osteoarthritis of the Knee: A Randomized Controlled Trial', *Annals of Internal Medicine*, 139 (2003), BBC, 724–30.
26. Colapinto, op. cit.
27. Celia Hatton, 'Nobel Prize Winner Tu Youyou Helped by Ancient Chinese Remedy', 6 October 2015.
28. Jane Perlez, 'Answering an Appeal by Mao Led Tu Youyou, a Chinese Scientist, to a Nobel Prize', *New York Times*, 6 October 2015.
29. 'The Nobel Prize in Physiology or Medicine 2015', Nobel press release, 5 October 2015.
30. Coué, 40.
31. Robertson, 30.
32. Massimo Pigliucci, 'How to be a Stoic', *New York Times*, 2 February 2015.
33. Robertson, 8.
34. Ibid., 166.
35. Kenneth E. Vail III et al., 'When Death is Good for Life: Considering the Positive Trajectories of Terror Management', *Personality and Social Psychology Review*, 16, no. 4 (2012), 303–29, cited in Kerr, 146.
36. Robertson, 250.
37. Nietzsche, 9.
38. Epictetus, xx.

3: The Missing Piece

1. Livio, 159.
2. Ken Rosenthal, 'US Men Czech Out, without Checking In', *Baltimore Sun*, 2 February 1998.
3. Rachel Alexander, 'Czech Republic Beats Russia for Gold', *Washington Post*, 23 February 1998.
4. John Henderson, 'Boy from the Black Sea: Vladimir Kramnik in Interview', *The Week In Chess*, 313 (6 November 2000).
5. Mark Crowther, 'Kramnik Takes Kasparov's World Title', *The Week In Chess*, 313 (6 November 2000).
6. Eric Schiller, *Learn from Kasparov's Greatest Games* (Las Vegas, 2005).
7. Garry Kasparov, 'Garry's Choice', *Chess Informant*, 118 (November 2013).
8. See the excellent biographical and documentary resources of the website devoted to Lamarck by the Centre national de la recherche scientifique: lamarck.cnrs.fr.
9. Christoph Marty, 'Darwin, Cuvier and Lamarck', *Scientific American*, 12 February 2009.
10. Schrödinger, 41.
11. Livio, 53.
12. Cited in Gould, 156.
13. Smith, 166.
14. Robin Marantz Henig, 'How Depressed Is That Mouse?', *Scientific American*, 7 March 2012.
15. Johnson, 205.
16. Helen Thomson, 'Study of Holocaust Survivors Finds Trauma Passed on to Children's Genes', *Guardian*, 21 August 2015.
17. Weismann, vol. II, 107–12.
18. Freud, 97.
19. Dawkins, 164–5.
20. Ayres, 90–96.

4: Game Changers

1. Emma Jay, 'Viagra and Other Drugs Discovered by Accident', BBC, 20 January 2010.
2. Davin Hiskey, 'The Shocking Story behind Playdoh's Original Purpose', *Business Insider*, 21 September 2015.
3. Sun Tzu, IX 15.
4. Ibid., III 2.
5. Freedman, 51.
6. Ibid., 509.
7. Sun Tzu, III 18.
8. Golitsyn, 40–1; Albats, 170.
9. Sun Tzu, XIII 21.
10. Golitsyn, 43.
11. Sun Tzu, I 22.
12. Ibid., V 18.

13. Golitsyn, 278–9.
14. Andrew & Mitrokhin, 528.
15. Epstein, 51.
16. Andrew & Mitrokhin, 479.
17. Epstein, 85.
18. Ibid., 16.
19. Ibid., 223.
20. Andrew & Mitrokhin, 233.
21. Walter Pincus, 'Yuri I. Nosenko, 81; KGB Agent Who Defected to the US', *Washington Post*, 27 August 2008.
22. Andrew & Mitrokhin, 241–3; 479.
23. Bacon, 127.
24. Aubrey, 75.
25. Janet Maslin, 'The Inventor Who Put Frozen Peas on Our Tables', *New York Times*, 25 April 2012.
26. Smith, Andrew F., 51.
27. Benedict Carey, 'LSD, Reconsidered for Therapy', *New York Times*, 3 March 2014.
28. Grof & Halifax, 15.
29. Erowid, 'From the Stolaroff Collection: Laura Huxley's Letter on Aldous' Passing', erowid.org, July 2009.
30. *CIA v. Sims*, 471 US 159 – Supreme Court 1985.
31. Ed Cumming, 'Is LSD about to Return to Polite Society?', *Observer*, 26 April 2015.
32. Teri S. Krebs & Pål-Ørjan Johansen, 'Lysergic Acid Diethylamide (LSD) for Alcoholism: Meta-analysis of Randomized Controlled Trials', *Journal of Psychopharmacology*, vol. 26, no. 7 (July 2012), 994–1002.
33. Matthew W. Johnson et al., 'Pilot Study of the 5–HT2AR Agonist Psilocybin in the Treatment of Tobacco Addiction', *Journal of Psychopharmacology*, vol. 28, no. 11 (November 2014), 983–92.
34. Peter S. Hendricks et al., 'Classic Psychedelic Use is Associated with Reduced Psychological Distress and Suicidality in the United States Adult Population', *Journal of Psychopharmacology*, vol. 29, no. 3 (March 2015), 280–8.
35. Suresh D. Muthukumaraswamy et al., 'Broadband Cortical Desynchronization Underlies the Human Psychedelic State', *Journal of Neuroscience*, vol. 33 no. 38 (8 September 2013), 15171–83.
36. Cumming, op. cit.
37. Dan Hurley, 'The Return of Electroshock Therapy', *Atlantic*, December 2015.

5: Are We Nearly There Yet?

1. Barbara Demick, 'A High-tech Approach to Getting a Nicotine Fix', *LA Times*, 25 April 2009.
2. 'E-cigarettes: an Evidence Update', PHE publications gateway number: 2015260.
3. 'Professor David Nutt Discusses e-cigarettes', National Institute for Health Innovation, University of Auckland, 8 December 2013.

4. James Dunworth, 'An Interview with the Inventor of the Electronic Cigarette, Herbert A. Gilbert', ecigarettedirect.co.uk, 2 October 2013.
5. 'Smokeless Non-tobacco Cigarette', US patent no. 3200819 A.
6. Robert N. Proctor, 'The History of the Discovery of the Cigarette–Lung Cancer Link: Evidentiary Traditions, Corporate Denial, Global Toll', *Tobacco Control*, 21 (2012), 87–91.
7. Diogenes, 36–8.
8. Lederman, 59.
9. Bohr, 13.
10. Kuhn, 133.
11. Weinberg, S., 13.
12. Bohr, 70.
13. Holton, 82.
14. Ball (2006), 10.
15. Ibid.
16. Shane Hickey, 'Are Britain's Foodies Ready to Eat Insects?', *Guardian*, 1 February 2015.
17. Ibid.
18. Marcel Dicke, 'Nordic Food Lab to Serve Insect Snacks at First Global Conference on Edible Insects', Nordic Food Lab press release, 15 May 2014.
19. grubkitchen.co.uk.
20. Emily Anthes, 'How Insects Could Feed the World', *Guardian*, 30 October 2014.
21. Ibid.
22. 'Ants and a Chimp Stick', Nordic Food Lab, 28 June 2013.
23. Anthes, op. cit.
24. Emma Bryce, 'Foodies Unite: Insects Should Be More Food than Fad', *Guardian*, 20 May 2014.
25. *Edible Insects: Future Prospects for Food and Feed Security*, FAO Forestry Paper 171, Food and Agriculture Organization of the United Nations, 2013.
26. Beyer, 242.
27. Ibid., 264.
28. Ibid., 273.
29. Philip Schieber, 'The Wit and Wisdom of Grace Hopper', *OCLC Newsletter*, no. 167 (March/April, 1987).
30. Beyer, 301.
31. Weinberg, G., 111.
32. Ensmenger, 236.
33. Julia Carrie Wong, 'Women Considered better Coders – but Only if they Hide their Gender', *Guardian*, 12 February 2016.

6: Something New under the Sun(s)

1. Chaucer, 385.
2. Quoted in Wootton, 75.

3. Ibid., 207.
4. Ibid., 345.
5. J. E. McGuire & P. M. Rattansi, 'Newton and the "Pipes of Pan"', *Notes and Records of the Royal Society of London*, vol. 21, no. 2 (December 1966), 108–43.
6. Wootton, 108–9.
7. Hugo Gernsback, 'The Isolator', *Science and Invention*, July 1925.
8. Thanks to Carl Cederström.
9. Freedman, 126.
10. Ibid., 129–31.
11. Ibid., 146.
12. Ibid.
13. Alexander J. Field, 'Schelling, von Neumann, and the Event That Didn't Occur', *Games,* 5 (2014), 53–89.
14. Freedman, 172–3.
15. Jean-Paul Carvalho, 'An Interview with Thomas Schelling', *Oxonomics* 2 (2007), 1–8.
16. Thomas C. Schelling, 'An Essay on Bargaining', *American Economic Review*, vol. 46, no. 3 (June 1956), 281–306.
17. Schelling, 45.
18. James Rothwell & Rob Crilly, 'Nato Says North Korea's "hydrogen bomb" Test "undermines international security"', *Telegraph*, 6 January 2016.
19. *Today*, BBC Radio 4, 6 January 2016.
20. Hans M. Kristensen, 'General Confirms Enhanced Targeting Capabilities of B61-12 Nuclear Bomb', Federation of American Scientists blogpost, 23 January 2014.
21. William J. Broad & David E. Sanger, 'As US Modernizes Nuclear Weapons, "Smaller" Leaves Some Uneasy', *New York Times*, 11 January 2016.
22. William J. Perry & Andy Weber, 'Mr. President, Kill the New Cruise Missile', *Washington Post,* 15 October 2015.
23. Melissa Locker, 'Stephen Hawking Has Finally Weighed in on Zayn Malik Leaving One Direction', *Vanity Fair*, 26 April 2015.
24. Hoyle (1960), 119.
25. Pius XII, 'The Proofs for the Existence of God in the Light of Modern Natural Science', Address to the Pontifical Academy of Sciences, 22 November 1951.
26. Livio, 201, 211.
27. Andrei Linde, 'A Brief History of the Multiverse', arXiv:1512.01203.
28. Wilczek (2009), 182.
29. George Ellis & Joe Silk, 'Scientific Method: Defend the Integrity of Physics', *Nature*, 16 December 2014.
30. Lewis, 2.
31. Ibid., 110.
32. Ibid., 2.
33. Ibid., 3.
34. Wilczek, 318–19.
35. *New Books In Philosophy* podcast, 'Margaret Morrison, "Reconstructing Reality: Models, Mathematics, and Simulations"', 15 July 2015.

7: The Jury's Still Out

1. Kondo, 95.
2. Ibid., 96.
3. Ibid., 86.
4. Ibid., 90.
5. Ibid., 70.
6. Ibid., 102.
7. Lucretius, trans. Martin Ferguson Smith, *De rerum natura* (Indianapolis, 2001), cited by Catherine Wilson, 'Commentary on Galen Strawson', in Strawson et al., 177.
8. Galen Strawson, 'Real Naturalism', *London Review of Books*, 26 September 2013.
9. Galen Strawson, 'Consciousness Myth', *TLS*, 25 February 2015.
10. Eddington, 258.
11. Chomsky, 142.
12. David Graeber, 'What's the Point if We Can't Have Fun?', *The Baffler*, no. 24 (2014).
13. Strawson et al., 248.
14. Skrbina, 88.
15. Schrödinger, 87.
16. Simon Blackburn, 'Thomas Nagel: a Philosopher Who Confesses to Finding Things Bewildering', *New Statesman*, 8 November 2012.
17. Mark Vernon, 'The Most Despised Science Book of 2012 is . . . Worth Reading', *Guardian*, 4 January 2013.
18. MacIntyre, 65.
19. Nagel, 28.
20. Ibid., 52.
21. Ibid., 6–7.
22. Ibid., 92.
23. John Hawthorne & Daniel Nolan, 'What Would Teleological Causation Be?', in Hawthorne, 265–84.
24. Galen Strawson, 'Panpsychism? Reply to Commentators with a Celebration of Descartes', in Strawson et al., 184–5.
25. Schopenhauer, I xvii.
26. Malthus, 148–9.
27. Steven Poole, 'Why Are We So Obsessed with the Pursuit of Authenticity?', *New Statesman*, 7 March 2013.

8: When Zombies Attack

1. Wootton, 120.
2. enclosedworld.com.
3. José Santiago, '36 Best Quotes from Davos 2016', World Economic Forum website, 23 January 2016.
4. Quiggin, 35.

5. Ibid., 64.
6. Ibid., 36.
7. Ibid., 168.
8. Firestein, 24.
9. Ibid., 24.
10. Edgerton, 156–7.
11. Amelia Gentleman, 'UK Firms Must Show Proof They Have No Links to Slave Labour under New Rules', *Guardian*, 28 October 2015.
12. Edgerton, 182.
13. Brad Bachtel et al., 'Polar Route Operations', *Aero* magazine (Boeing), QTR_04 2001.
14. Heath & Heath, 5.
15. Cass R. Sunstein, 'Believing False Rumors', in Levmore & Nussbaum, 106.
16. 'Ukraine Health Officials Fear Big Polio Outbreak', BBC, 22 September 2015.
17. Daniel D'Addario, 'The Music World's Fake Illuminati', *Salon*, 24 January 2013.
18. Suzanne Goldenberg, 'Work of Prominent Climate Change Denier was Funded by Energy Industry', *Guardian*, 21 February 2015; Douglas Fischer, '"Dark Money" Funds Climate Change Denial Effort', *Scientific American*, 23 December 2013; etc.
19. See Michael Dutton, Hsiu-ju Stacy Lo & Dong Dong Wu, *Beijing Time* (Cambridge, Mass., 2008).
20. Wilczek (2009), 6.
21. Helmholtz (1895), 229.
22. Aileen Fyfe, 'Peer Review: Not as Old as You Might Think', *THES*, 25 June 2015.
23. John P. A. Ioannidis, 'An Epidemic of False Claims', *Scientific American*, 17 May 2011.
24. Ed Yong, 'Nobel Laureate Challenges Psychologists To Clean Up Their Act', *Nature*, 3 October 2012.
25. Gary Gutting, 'Psyching Us Out: The Promises of "Priming"', *New York Times*, 31 October 2013.
26. 'How Science Goes Wrong', *Economist*, 19 October 2013.
27. John P. A. Ioannidis, 'Why Most Published Research Findings Are False', *PLoS* 10.1371/journal.pmed.0020124, 30 August 2005.
28. Ben Goldacre, 'Scientists Are Hoarding Data and it's Ruining Medical Research', *Buzzfeed*, 23 July 2015.
29. 'How Science Goes Wrong', *Economist*, 19 October 2013.
30. John Colapinto, 'Material Question', *New Yorker*, 22 December 2014.
31. Kuhn, 150–1.
32. Joseph Nocera, 'The Heresy That Made Them Rich', *New York Times*, 29 October 2005.

9: How to Be Wrong

1. Elizabeth Howell, '"Clever Editing" Warps Scientists' Words in New Geocentrism Film', *Live Science*, 15 April 2014.
2. 'Brahe Myths Are Disproved, but Secret Remains Buried', *Copenhagen Post*, 16 November 2012.
3. Wootton, 12–13.

4. Dennis Danielson & Christopher M. Graney, 'The Case against Copernicus', *Scientific American*, January 2014.
5. Danielson & Graney, op. cit.
6. Grant, 345.
7. Wootton, 193.
8. Kuhn, 79.
9. Bohr, 56.
10. *OED*.
11. Brooks (2009), 46–55.
12. John Maddox, 'A Book for Burning?', *Nature*, 24 September 1981.
13. Rupert Sheldrake, 'A New Science of Life', *New Scientist*, 18 June 1981.
14. D.J. Bem, 'Feeling the Future: Experimental Evidence for Anomalous Retroactive Influences on Cognition and Affect', *Journal of Personal and Social Psychology*, vol. 100, no. 3 (March 2011), 407–25.
15. Benedict Carey, 'Journal's Paper on ESP Expected to Prompt Outrage', *New York Times*, 5 January 2011.
16. Ned Potter, 'ESP Study Gets Published in Scientific Journal', ABC News, 6 January 2011.
17. Carey, op. cit.
18. Daryl Bem, Patrizio Tressoldi, Thomas Rabeyron & Michael Duggan, 'Feeling the Future: A Meta-analysis of 90 Experiments on the Anomalous Anticipation of Random Future Events', *F1000Research*, 4 (2015), 1188.
19. E.J. Wagenmakers, 'Bem is Back: A Skeptic's Review of a Meta-Analysis on Psi', Open Science Collaboration, 25 June 2014.
20. Edgerton, 210.
21. Crick, 99.
22. Gould, 96.
23. Ewen Callaway, 'Fearful Memories Haunt Mouse Descendants', *Nature*, 1 December 2013.
24. Scott F. Gilbert, John M. Opitz & Rudolf A. Raff, 'Resynthesizing Evolutionary and Developmental Biology', *Developmental Biology*, 173 (1996), 357–72, 365.
25. Ibid., 366.
26. Ellen Larsen, 'Genes, Cell Behavior, and the Evolution of Form', in Gerd B. Müller & Stuart A. Newman (eds), *Origination of Organismal Form: Beyond the Gene in Developmental and Evolutionary Biology* (Cambridge, Mass., 2003), 125.
27. Marc Kirschner, John Gerhart and Tim Mitchison, 'Molecular "Vitalism"', *Cell*, vol. 100 (7 January 2000), 79–88.
28. Timothy O'Connor & Hong Yu Wong, 'Emergent Properties', *Stanford Encyclopedia of Philosophy* (Summer 2015); Scott F. Gilbert & Sahotra Sarkar, 'Embracing Complexity: Organicism for the 21st Century', *Developmental Dynamics*, 219 (2000), 1–9.
29. Livio, 241.
30. Wilczek (2015), 318.
31. Lightman, 8.
32. Montaigne, 643.
33. Kuhn, 171.

34. Firestein, 15–16.
35. Ibid., 5.
36. Loewenstein, George, 'The Psychology of Curiosity: A Review and Reinterpretation', *Psychological Bulletin*, vol. 116, no. 1 (1994), 87.
37. Ibid., 91.
38. Ibid., 94.
39. Feyerabend (1988), 157.
40. Ibid., 27.
41. Ibid., 116.

10: The Placebo Effect

1. See Steven Poole, 'Your Brain on Pseudoscience: the Rise of Popular Neurobollocks', *New Statesman*, 6 September 2012.
2. Gabrielle Glaser, 'The Irrationality of Alcoholics Anonymous', *Atlantic*, April 2015.
3. Melissa Davey, 'Marc Lewis: the Neuroscientist Who Believes Addiction is not a Disease', *Guardian*, 30 August 2015.
4. Helmholtz (1875), 181.
5. Helmholtz (1925), III 4–5.
6. Cited in Brent, 72.
7. Capri, 88.
8. Jo Marchant, 'Strong Placebo Response Thwarts Painkiller Trials', *Nature*, 6 October 2015.
9. Asbjørn Hróbjartsson & Peter C. Gøtzsche, 'Is the Placebo Powerless? An Analysis of Clinical Trials Comparing Placebo with No Treatment', *New England Journal of Medicine*, 344 (24 May 2001), 1594–602.
10. Luana Colloca & Fabrizio Benedetti, 'Placebos and Painkillers: Is Mind as Real as Matter?', *Nature Reviews: Neuroscience*, vol. 6 (July 2005), 545–52.
11. Brooks (2009), 170.
12. Benedetti, 195.
13. Brooks (2009), 175.
14. Plato, *Charmides* (157a–157b), 7.
15. Fabrizio Benedetti, 'How Placebos Change the Patient's Brain', *Neuropsychopharmacology*, 36 (1) (January 2011), 339–54.
16. Brooks (2009), 167.
17. T. J. Kaptchuk et al., 'Placebos without Deception: A Randomized Controlled Trial in Irritable Bowel Syndrome', *PLoS ONE 5* (12), e15591.
18. Coué, 5.
19. Ibid.
20. Luana Colloca & Fabrizio Benedetti, 'Placebos and Painkillers: Is Mind as Real as Matter?' *Nature Reviews: Neuroscience*, vol. 6 (July 2005), 551.
21. Oliver Burkeman, 'Therapy Wars: The Revenge of Freud', *Guardian*, 7 January 2016.
22. Nietzsche, 4–5.
23. James, 88.

24. Burns (2012), fig. 7–5.
25. Burns (1999), 113.
26. Wiseman, 11, 88–9, 244.
27. Leonard Cohen, 'That Don't Make It Junk'.
28. Wiseman, 11.
29. S. Schachter & J. Singer, 'Cognitive, Social, and Physiological Determinants of Emotional State', *Psychological Review*, 69 (1962), 379–99.
30. Kate Murphy, 'The Right Stance Can Be Reassuring', *New York Times*, 3 May 2013.
31. Jones, 242.
32. See, e.g., Tim Crane, 'Ready or Not', *TLS*, 14 January 2005; Tim Bayne, 'Libet and the Case for Free Will Scepticism', in Richard Swinburne (ed.), *Free Will and Modern Science* (Oxford, 2011).
33. Searle, 211.
34. Daniel Dennett, 'Reflections on "Free Will"', naturalism.org, 24 January 2014.
35. Kathleen D. Vohs & Jonathan W. Schooler, 'The Value of Believing in Free Will: Encouraging a Belief in Determinism Increases Cheating', *Psychological Science*, vol. 19, no. 1 (January 2008), 49–54.
36. David Bourget & David J. Chalmers, 'What Do Philosophers Believe?' *Philosophical Studies*, vol. 170, no. 3 (2014), 465–500.
37. Schrödinger, 89.
38. Dyson, 249.

11: Utopia Redux

1. Marcusc, 154.
2. Tett, 132.
3. Ibid.
4. Thanks to Tony Yates.
5. Keynes, 142, 341.
6. Paul Krugman, 'Conservatives and Keynes', *New York Times*, 20 May 2015.
7. Ibid.
8. Paul Krugman, 'The Smith/Klein/Kalecki Theory of Austerity', *New York Times*, 16 May 2013.
9. Jacques Derrida, trans. Peggy Kamuf, *Spectres of Marx* (1994; London, 2006), 10
10. Malthus, 4.
11. Marcuse, 4.
12. Thomas Paine, 'Agrarian Justice, Opposed to Agrarian Law, and to Agrarian Monopoly, Being a Plan for Meliorating the Condition of Man, &c.', *Eighteenth Century Collections Online*, University of Michigan.
13. John Cunliffe and Guido Erreygers, 'The Enigmatic Legacy of Charles Fourier: Joseph Charlier and Basic Income', *History of Political Economy*, vol. 33, no. 3 (Fall 2001).
14. Ibid.
15. Enno Schmidt & Paul Solman, 'How a "stupid painter from Switzerland" is Revolutionizing Work', *PBS Newshour*, 9 April 2014.

16. Lauren Smiley, 'Silicon Valley's Basic Income Bromance', *Backchannel*, 15 December 2015.
17. Evelyn Forget, 'The Town with No Poverty: The Health Effects of a Canadian Guaranteed Annual Income Field Experiment', *Canadian Public Policy*, vol. 37, no. 3 (2011) 283–305.
18. Farhad Manjoo, 'A Plan in Case Robots Take the Jobs: Give Everyone a Paycheck', *New York Times*, 2 March 2016.
19. 'Pennies from Heaven', *Economist*, 26 October 2013.
20. See Poole, 16–19.
21. Cited in Michael Millgate, *Thomas Hardy: A Biography* (Oxford, 1982), 266.
22. Graeber, 177.
23. Alexander Guerrero, 'The Lottocracy', *Aeon*, 23 January 2014.
24. Scruton, 62.
25. Guerrero, op. cit.
26. Stiglitz, 302.

12: Beyond Good and Evil

1. William Gibson, 'Talk of the Nation', National Public Radio, 30 November 1999.
2. Edwin Black, 'Eugenics and the Nazis – the California Connection', *San Francisco Chronicle*, 9 November 2003.
3. Jonathan Freedland, 'Eugenics: The Skeleton That Rattles Loudest in the Left's Closet', *Guardian*, 17 February 2012.
4. Alexander G. Bassuk et al., 'Precision Medicine: Genetic Repair of Retinitis Pigmentosa in Patient-derived Stem Cells', *Nature Scientific Reports* 6, Article number 19969 (2016).
5. Galton, 18 n. 2.
6. Ibid., 200.
7. Ibid., 199.
8. Ibid., 201.
9. Ibid., 202.
10. Ibid., 204.
11. Ibid., 214.
12. MacKellar & Bechtel, 191.
13. Habermas, 60–1.
14. 'God's little helper', *THES*, 14 March 1997.
15. Andrew J. Imparato & Anne C. Sommers, 'Haunting Echoes of Eugenics', *Washington Post*, 20 May 2007.
16. Basil H. Aboul-Enein, 'Preventive Nutrition in Nazi Germany: A Public Health Commentary', *Online Journal of Health Ethics*, vol. 9, issue 1.
17. James J. H. Rucker, 'Psychedelic Drugs should be Legally Reclassified so that Researchers can Investigate their Therapeutic Potential', *BMJ*, vol. 350, h2902 (26 May, 2015).
18. 'The War on Drugs at a Glance', leap.cc.
19. Chris Roberts, 'Black City Residents More Likely to be Busted for Pot, ACLU Says', *San Francisco Examiner*, 6 June 2013.

20. D. J. Nutt, 'Equasy: An Overlooked Addiction with Implications for the Current Debate on Drug Harms', *Journal of Psychopharmacology*, 23 (1) (2009), 3–5.
21. Robert J. Blendon & John T. Young, 'The Public and the War on Illicit Drugs', *Journal of the American Medical Association*, vol. 279, no. 11 (18 March 1998), p. 827.
22. 'America's New Drug Policy Landscape', PewResearchCenter, 2 April 2014.
23. 'Majority Now Supports Legalizing Marijuana', PewResearchCenter, 4 April 2013.
24. 'Depression', Fact sheet no. 369, World Health Organization, October 2015.
25. Morozov, 171.
26. Gary Marcus, 'Moral Machines', *New Yorker*, 24 November 2012.
27. Wolstenholme, 288.
28. Ibid., 275.
29. Ibid., 283.
30. Ibid., 275.
31. Petersen, 69–71.
32. Malthus, 10.
33. Ibid., 22–3.
34. Ibid., 23.
35. Petersen, 71.
36. Malthus, 30.
37. Letters, *New Yorker*, 21 September 2015.
38. Petersen, 58.
39. Cited ibid., 219.
40. Allan, 58.
41. Zhuang Pinghui, 'Beijing Unfazed by Drop in Births despite Ending the One-child Policy', *South China Morning Post*, 21 January 2016.
42. Sarah Conly, 'Here's Why China's One-child Policy was a Good Thing', *Boston Globe*, 31 October 2015.

13: Don't Start Believing

1. Diogenes, 62.
2. Ibid., 60.
3. Ibid., 59.
4. Adamson, 102.
5. Ibid., 123.
6. Diogenes, 68.
7. Ibid., 66–7.
8. Ibid., 59.
9. Caryle Murphy, 'Most Americans believe in Heaven . . . and Hell', PewResearchCenter, 10 November 2015.
10. Unger, 221–2.
11. Ibid., 28 n. 3.
12. Ibid., 238 n. 14.
13. Heuer, 76.

14. Epstein, 75.
15. N. David Mermin, 'Could Feynman Have Said This?', *Physics Today*, May 2004, 10.
16. Levitt & Dubner, 88.
17. Firestein, 21.
18. Heuer, 104.
19. Brooks (2009), 187.
20. Yuki Nagata, Kota Usui & Mischa Bonn, 'Molecular Mechanism of Water Evaporation', *Physical Review Letters,* vol. 115, no. 23 (4 December 2015), article 236102.
21. Brooks (2009), 190.
22. Ibid., 200.
23. Ibid., 200.
24. Heuer, 74–5.
25. Steven Poole, 'Not So Foolish', *Aeon*, 22 September 2014.
26. Heuer, 109.

Epilogue: Back to the Future

1. 'Science on the Surface of a Comet', European Space Agency press release, 30 July 2015.
2. 'First Detection of Molecular Oxygen at a Comet', European Space Agency press release, 28 October 2015.
3. 'New Findings from NASA's New Horizons Shape Understanding of Pluto and its Moons', Nasa, 17 December 2015.
4. 'New Horizons Finds Blue Skies and Water Ice on Pluto', Nasa, 8 October 2015.
5. Vance, 226.
6. Ken Kremer, 'A City on Mars is Elon Musk's Ultimate Goal Enabled by Rocket Reuse Technology', *Universe Today*, 27 December 2015.
7. Vance, 336.
8. Ibid., 249.
9. 'Target Earth', Near Earth Object Program, nasa.gov.
10. 'Russia's Improved Ballistic Missiles to be Tested as Asteroid Killers', Tass, 11 February 2016.
11. 'Project Overview: NEOShield-2: Science and Technology for Near-Earth Object Impact Prevention', neoshield.net; 'DE-STAR: Directed Energy Planetary Defense', deepspace.ucsb.edu.
12. Matt Vella, 'Yuri Milner: Why I Funded the Largest Search for Alien Intelligence Ever', *Time*, 20 July 2015.
13. Eric Berger, 'NASA's Newest Cargo Spacecraft Began Life as a Soviet Space Plane', *Ars Technica*, 18 January 2016.
14. John Ruwitch, 'China to Land Probe on Dark Side of Moon in 2018: Xinhua', Reuters, 15 January 2016.
15. *Nasa's Journey to Mars: Pioneering Next Steps in Space Exploration*, Nasa, 8 October 2015.

Index

INDEX

INDEX

INDEX

INDEX